C. M. Knop

INTERFERENCE SUPPRESSION TECHNIQUES for MICROWAVE ANTENNAS and TRANSMITTERS

INTERFERENCE SUPPRESSION TECHNIQUES for MICROWAVE ANTENNAS and TRANSMITTERS

Ernest R. Freeman

ARTECH HOUSE

To my wife, June, and Jesse, Miriam, Sarah and Beth
for their forebearance and understanding

ARTECH HOUSE, INC.
610 Washington St., Dedham, MA

International Standard Book Number: 0-89006-110-6
Library of Congress Catalog Card Number: 81-71049

CONTENTS

C. M. Knop

INTRODUCTION

The subject of interference suppression requires a knowledge of all the areas contributing to the problem: the generators of desired and undesired signals, the coupling paths between them and the receiving systems, and the receiving systems themselves, along with their information processing and decision making functions. There has been a great deal of effort devoted to modelling these areas in order to predict interference levels. Once such areas have been evaluated the question of what to do about the interference becomes paramount.

Our objective is to provide information that will present appropriate techniques to enable systems engineers and designers to meet EMC specifications and standards for high power microwave systems. The areas covered are antenna and site treatments and transmitter techniques.

The use of innovative antenna techniques for the purpose of reducing off-frequency and off-boresight radiation and reception has been the subject of many widely dispersed investigations. In many cases the techniques were studied in order to increase the on-frequency, on-boresight performance without specific regard for the interference. In other cases, such techniques have been the result of efforts directed towards other concerns but also applicable to interference suppression. In such instances, the antenna and filter techniques are usually documented, if at all, in places that are not easily cross-referenced to the interference suppression area.

ANTENNA TECHNIQUES

This book presents appropriate antenna and transmitter techniques of this nature, emphasizing those that can be used to retrofit existing designs. Flat plate array, the conical scan parabolic, monopulse or conical scan parabolic, monopulse or conical scan receive only (COSRO) fed parabolic, Cassegrainian and compound (Cassegrain with null feed horn) antenna systems, and pillbox and array antennas are discussed herein. The antenna effort was primarily directed at determining the improvement in antenna pattern characteristics that could be obtained by the use of the technique, assuming that the mechanical drive systems of the antennas could also be modified, where required, within reason. It is understood that many of these drive systems are designed to very precise safety factors and that almost any additional weight added to the antenna will necessitate some drive modifications.

The design of an antenna is a function of the mission or purpose for which the overall system is intended. The characteristics of an antenna are usually specified in terms of its terminal impedance, bandwidth, power gain, polarization, pattern, and scanning characteristics. The most important of these characteristics in regard to electromagnetic compatibility are the bandwidth and pattern, although the other factors are also significant. It might be more descriptive to say that the characteristics which are most amenable to change for the purpose of achieving compatibility are the bandwidth and out-of-the-main-lobe pattern.

All the mentioned characteristics are usually specified and controlled within the design band. It is extremely unusual to find limits on out-of-band performance. The fact that transmitters and receivers emit and respond to energy which is out-of-band and off-boresight makes consideration of such parameters essential for compatible operation.

The techniques used to make aperture antennas more selective, both from a frequency and spatial standpoint, can be classified as:

1. Site Modifiers
   a) Clutter fences
   b) Removal of obstacles, etc.

2. Antenna Configuration Modifiers
    a) Hooded antennas
    b) Tunnel antennas
    c) Selective placement of radar absorbing material
    d) Dielguide antennas
    e) Extensions of reflector areas, etc.
3. Signal Processing Modifiers
    a) Side-lobe blanking and cancelling
    b) Adaptive array techniques, etc.

This book will emphasize the first two areas.

The sources of significant undesired types of radiation may best be discussed using an example. An aperture antenna such as a paraboloidal reflector may be characterized by an antenna pattern consisting of three major regions. An analysis of the radiation of the aperture proper is descriptive of the radiation only in the vicinity of the main beam and the first few sidelobes. Most of the research on the radiation patterns of antennas has been concerned with this region. The other areas are dominated by the effects of direct feed radiation and energy diffracted at the aperture edges. It can be demonstrated, and has been noted by several observers, that main beam interference is not the significant source of interference problems, but that the other areas under consideration are far more important. This is especially true in the case of rotating surveillance radars where the main beam is seen less than 1 percent of the time, and in the case of space systems where the main beams are directed skywards.

In the case of the paraboloidal reflector, the back lobe radiation is caused by two major contributors:

(1) Edge diffraction around the rim of the reflector, which can be treated by the geometrical theory of diffraction; and
(2) Direct feed through radiation from the feed, which is not intercepted by the reflector proper.

The edge diffraction problem has been attacked by the application of radar absorber material (RAM), or the use of small flanges around the aperture edges. These methods have been demonstrated experimentally [1]. The direct feed-through problem has been approached through redesign of the feed or by directly intercepting this energy. This can be done by extending the reflector area with inexpensive mesh material [2], or by the use of absorbing material placed outside the antenna proper. Redesign of the feed has reduced radiation to 45 dB below the main beam for angles greater than 30°, and 60 dB down for angles greater than 90° (or the limitations of the measuring equipment in certain cases) [3]. The tunnel antenna method for reducing direct-feed radiation has been investigated and produced similar results [4].

ANTENNA PATTERN CONSIDERATION IN EMC

The effect of high-gain scanning antennas on interference analysis is to require consideration of continuously changing levels of signals due to the relative motion of the antennas involved. This is most commonly done by consideration of the mutual gain of the antennas as a statistic. In other words, the one-transmitter-to-one-receiver interference situation can be modeled as

$$P_r = P_T G_T G_r G_L$$

where P_r represents the received power, $G_T G_r$ represents the mutual antenna gain, and G_L represents the propagation medium gain. It has been previously stated that the main beam gain would be observed less than 1% of the time for a typical radar antenna. It has also been shown many times [2] that the out-of-mainbeam antenna pattern may be represented by a normal distribution. If both antennas in the situation of interest are of the scanning type then the mutual gain can be determined by the convolution of the individual antenna pattern distributions, which, if normal, is also a normal distribution whose mean is the sum of the means and whose standard deviation is the square root of the sum of the squares of the individual standard deviations.

Once the model is derived, interpretation of the model in terms of interference requires consideration of the specific systems involved. For example, rotating surveillance radars would pass through the distribution levels at a regular and predictable rate, e.g., five times per minute. In such a case, if a receiving system were susceptible to main beam interference energy only, such as a "PRF buzz" heard on AM radios, then its occurrence at such a regular rate may or may not be a significant problem. The important point is to realize that although on a probabilistic basis there would be less than a .01 probability of this type of occurrence, in actuality there is a certainty of it occurring a fixed percentage of the time. On the other hand, if the situation involves tracking antennas, whose look-angle is based on target position, the probabilistic representation of the mutual gain assumes an entirely different meaning, where the main-beam illumination condition may never occur; or it may occur, and continue for a considerable length of time. It is important, therefore, to assess each situation carefully before reaching conclusions on the basis of probabilistic mutual antenna gain considerations.

TABLE 1

Mean and Standard Deviations for Side and Back Lobe Gains of High Gain Antennas

Site Type	*Mean (dB/iso.)*	*Standard Deviation (dB)*
Open, or Isolated Installation	-15	8
Average Installation	-10	6
Crowded, Or Congested, Installation	- 5	4

The effect of applying the antenna techniques described herein, with regard to the probabilistic or statistical description of antenna gain, may be estimated by considering the results of past studies designed to determine the statistical description of the side and back lobe region of antennas. Table 1 is a group of estimators used in interference analysis, based on reduction of empirical data [5]. Note that, for crowded or congested installations such as may be found on a ship or an antenna farm, the effects of the signal being reflected from many adjacent objects causes the median gain of the side and back lobes to be higher and the standard deviation to be reduced, as would be expected. Now suppose that radar absorbing material (RAM) were used to absorb the energy that would normally impinge on these adjacent objects. The effect would be to make the congested installation appear as an open or isolated installation from the electromagnetic viewpoint. This could provide up to a 10 dB improvement in the mean side and back lobe suppression and a 4 dB increase in the standard deviation. The overall EMC effect is illustrated in Figure 1. In this example two interference thresholds are postulated, i.e., A

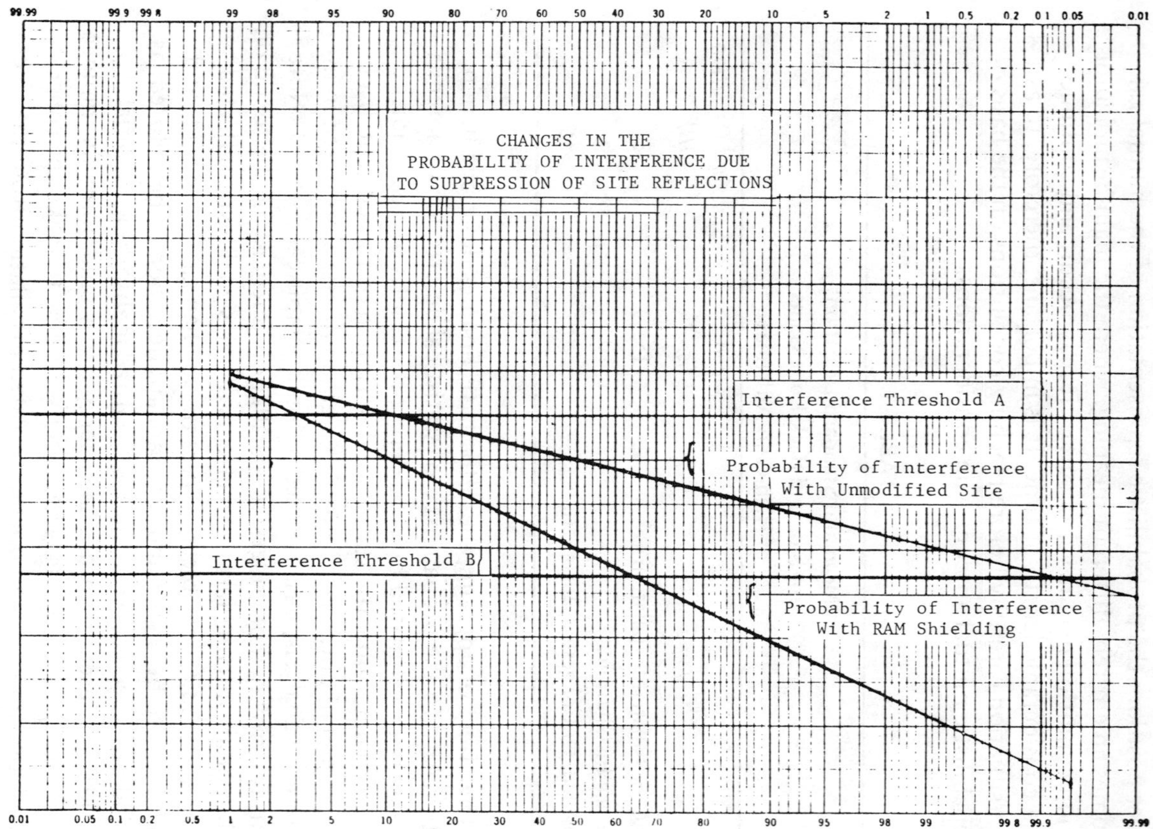

Figure 1 Effect of reducing site reflections

and B. For example, for A it is assumed that the probability of side and back lobe interference, with the antenna mounted in a congested location, is about 0.1. The application of RAM to make the site appear open would decrease this interference probability to 0.03. Suppose now that the interference threshold is that given by level B. This would result in a probability of 0.999 of having interference; an almost certain condition. The application of RAM to the scatterers on the site would lower this interference probability to 0.62, which may be a tolerable level.

The purpose of this example is to illustrate two major points. First, that from an interference standpoint, improving the antenna pattern can yield a considerable benefit over the general situation aspect and, further, that this benefit increases in value proportionally to the severity of the problem. Secondly, the estimation of improvement in a specific situation, based on mean or median gain alone, is not valid, but the overall statistics of the improvement in terms of distribution and standard deviations must be considered.

The effects due to the application of the RAM would be similar to the case of using tunnel antennas, which suppress side lobes. The important point is that the improvement to be attained is much greater on a probabilistic basis than may be apparent by looking at one specific path between a transmitting and receiving antenna. Another significant aspect is that the effects of the site of the antenna installation is at least as important in determining interference as the antenna pattern taken in isolation.

1

ABSORBERS

GENERAL DESCRIPTION OF RADAR ABSORBING MATERIALS

Historically, radar absorbing materials were first used in World War II to deny radar detection of German submarines. The Germans also proposed to use RAM to disguise large industrial plants from detection. Lately these materials have been applied to airplanes and missiles to reduce their radar cross sections. Anechoic chambers use large quantities of RAM in an attempt to provide a "free space" environment or one that is devoid of specular reflections.

More recently, RAM is being thought of as one way to reduce undesired coupling from nearby radar systems, such as those found aboard ship. Since most radar antenna patterns exhibit fairly strong sidelobes, which are an interference source for nearby systems, there is a need for a means for decoupling. Another source of interference for a radar may be a nearby structure which has a very strong reflection over a wide (for its size) range of angles. One way of reducing the reflection is to cover the structure with RAM. For example, aboard a ship there may be a tall whip antenna which represents a large source of interference to the nearby radar. Its radar image, or cross section, can be reduced by covering it with RAM material tuned to the frequency of the radar but transparent to the whip antenna's frequency of operation. Likewise, other offending surfaces aboard a ship may be covered with the appropriate RAM materials to reduce their radar cross sections. Another means for increasing the decoupling of two systems is through the use of tunnel antennas. In this usage, RAM is applied to the interior and edges of the tunnel to absorb the sidelobe energy.

There are basically two kinds of radar absorbing techniques in common use today. The first is the true absorber which converts incident energy into heat. The other is the multicore sandwich absorber which utilizes phase and amplitude cancellation of the incident and reflected wave at the surface of the sandwich. Since this is a narrowband device, several layers designed for different frequencies enable operation over a wide band of frequencies; this is also known as the Salisbury Screen.

Figure 1.1 "HAIR" type absorber

The most effective of the true absorbers are those based on the graded dielectric principle. The idea here is to make the change from "free space", or air, to a highly absorptive medium as gradual as possible. This is because any sharp discontinuity of the dielectric constant results in partial reflection of the incident wave. One of the first methods of providing a smooth transition from a lossless medium to a highly lossy one was the development of "Hair." The "Hair" type absorber is composed of enmeshed, rubberized, lossy fibers (see Figure 1.1).

The absorptive materials is usually carbon black. Because of structural requirements, a strong, lightweight absorber may be required. An example of such a type is the rigid plastic foam Type P absorber (Figure 1.2). This type of construction relies on many reflections of the wave within honeycombs to effect the attenuation from the lossy walls.

The most effective of the graded dielectric absorbers are the foam pyramid types (see Figure 1.3). The sharp point and gradual taper provide a smoother transition to the lossy material that can be achieved through use of a flat plate of graded dielectric. Since it is extremely difficult to make a flat surface which has a gradual change of dielectric constant, there always exists an "impedance step" of discontinuity which results in a reflection. The sharply pointed, pyramidal shape is a geometrical way of presenting to the

Figure 1.2 Rigid plastic form type P absorber

wave the equivalent of a graded dielectric material. The deeper the pyramid can be made, the better it will perform, because this allows more of the wave to be attenuated and serves to hide the discontinuity at the base.

One advantage of the multicore sandwich absorber or Salisbury Screen, as it is sometimes called, is that it is fairly thin and lightweight. It can be made rugged and flexible for general outdoor use, or it can be made structurally very strong and incorporated in planes or missiles to reduce their radar cross sections. It is often advertised by the manufacturers as a resonant foam absorber.

THEORY OF ABSORBERS

Dissipation in a Medium

Absorption of microwave energy in a dissipative medium is accomplished by converting the energy into heat. The propagation of electromagnetic energy and how different mediums affect this propagation are, therefore, of interest. For simple electromagnetic plane waves traveling in a dielectric medium or "free space", the electric and magnetic components are of the same form and are perpendicular to each other, and to the direction of travel. The magnitudes of the electric and magnetic fields are related in "free space" by the intrinsic impedance, n, which has a value of about 377 ohms. This relation can be expressed as:

$$E_y = \frac{\sqrt{\mu_o}}{\mathcal{E}_o} H_z \qquad \text{where the direction of propagation is in the X direction} \tag{1.1}$$

$$= nHz$$

μ_o = magnetic permeability

$\mathcal{E}_o$ = electric permittivity

When an electromagnetic wave travels in a substance, the values of permeability and permittivity are greater than, or at least equal to, their respective values in "free space."

One solution of the wave equation expresses the E-field as follows:

$$E_y = E_m e^{jwt \pm \Gamma x} \tag{1.2}$$

where: E_m is the magnitude of the E-field at the surface of the material

Γ = complex propagation constant

x = depth of penetration

If the propagation is in a perfect dielectric medium the conductivity, γ, is zero and $\Gamma = jw\sqrt{\mu\epsilon}$. In general, however, the conductivity, γ, of the material is included in the complex propagation constant as follows:

$$\Gamma = jw \quad \mu\epsilon \quad 1 + \frac{\gamma}{jwe} \tag{1.3}$$

$$= \alpha + j\beta$$

α = attenuation constant

β = phase constant

Equation (1) can now be written as:

$$E_y = E_m e^{j(wt - \beta x)} e^{-ax} \tag{1.4}$$

The factor e^{-ax} shows the attenuation with distance into the material. When the conductivity is finite but small, as in a good conductor, the wave travels far with little attenuation. But when the conductivity is high, as in metal, the wave is attenuated in a fraction of a millimeter. When the dielectric constant is complex, the intrinsic impedance is also complex as seen from:

$$n = \frac{\mu}{\epsilon(1 + \frac{\gamma}{j\omega\epsilon})} \tag{1.5}$$

The attenuation factor a can be found from a binomial-series expansion of Γ. For small values of conductivity $a = (\gamma/2)(\sqrt{\mu\epsilon}/\epsilon)$. This can be modified in the following manner.

$$\alpha = \frac{1}{2} \sqrt{\mu\epsilon} \; \frac{\gamma}{\omega\epsilon} \quad ; \; \lambda = \frac{2\pi c}{\omega} \qquad c = \text{speed of light} = \frac{1}{\sqrt{\mu_o \epsilon_o}} \tag{1.6}$$

$$= \frac{\pi c}{\lambda} \sqrt{\mu\epsilon} \; \frac{\gamma}{\omega\epsilon} \quad \text{nepers/meter}$$

The quantity $|\lambda/\omega\epsilon|$ is sometimes known as the power factor or loss tangent and is expressed as $\tan\delta = |\gamma/\omega\epsilon|$ so that

$$\alpha = \frac{\pi \tan\delta}{\lambda} \; \frac{\mu\epsilon}{\mu_o \epsilon_o} \quad \text{nepers/meter} \tag{1.7}$$

Power loss in decibels per meter is then:

$$8.686\,\alpha = \frac{8.686}{\lambda} \pi \tan\delta \; \frac{\mu\epsilon}{\mu_o\epsilon_o} \quad \text{for } 0.0001 \leqslant \tan\delta \leqslant 0.5 \tag{1.8}$$

Nomographs for attenuation in a dielectric are given in the section "Nomographs and Trade-off Curves," below.

As can be seen from Equations (1) and (2), the three quantities μ, ϵ, and γ affect the attenuation of an electromagnetic wave in a material. The wave is attenuated in direct proportion to the magnitudes of the three quantities. The easiest of the three quantities to control is conductivity, γ; most absorbers utilize this quantity, in the form of carbon black, to create a highly dissipative medium.

Figure 1.3 Typical use of a tapered pyramidal absorber

Reflection

When an electromagnetic wave traveling in one medium enters a second medium that has a different dielectric constant, a reflected wave is generated at the boundary. Since it is not possible to absorb that portion which is reflected, it is desirable to minimize this portion. It will be shown that reflection is a function of the complex intrinsic impedances of the two materials and of the angle of incidence.

The general case of a wave of any polarization, traveling in any medium, and entering another medium with some arbitrary orientation can become very complicated. However, the wave vector may be thought of as being comprised of components that are parallel or perpendicular to the reflecting surface. Reflection of a wave with parallel polarization will be analyzed; a similar approach may be taken for perpendicular polarization.

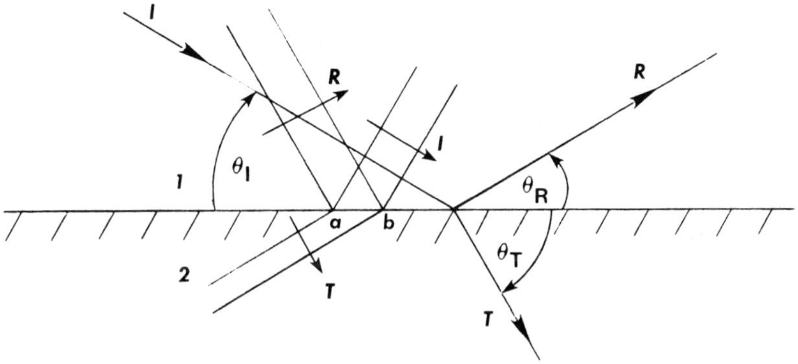

Figure 1.4 Relations between incident, reflected, and transmitted waves at the interface

In Figure 1.4, I is a wave traveling in region 1 that is incident on a surface separating regions 1 and 2. The reflected wave is R and the transmitted wave is T. The velocity of propagation is different in the two regions. The solution is based on the boundary condition requiring that the tangential components of the E- and H-waves be continuous.

$$H_{tI} + H_{tR} = H_{tT} \qquad t = \text{the tangential component} \tag{1.9}$$

Since the E-field is parallel to the plane of reflection:

$$E_I + E_R = E_T \qquad \text{(dropping t)} \tag{1.10}$$

For the above two equations to be satisfied, the three waves must have the same frequency, be moving in the same direction, and at the boundary they must have the same distance from wave-crest to wave-crest. This is the distance (a-b) in Figure 1.4.

If region 1 is air with a dielectric constant of ϵ_o and region 2 is a material with a dielectric constant of ϵ such that $\epsilon > \epsilon_o$, then the wave velocity is slower in region 2 than in air or "free space". This can be seen from the following:

$$c = \frac{1}{\sqrt{\mu_o \epsilon_o}} \qquad c = \text{the velocity of propagation in a vacuum} \tag{1.11}$$

$$v = \frac{1}{\sqrt{\mu \epsilon}} \qquad v = \text{the velocity of propagation in a substance} \tag{1.12}$$

Since μ and ϵ are greater than μ_o and ϵ_o, then $c > v$. The quantities μ and ϵ are the permeability and permittivity of the substance and always have greater values for a substance than for a vacuum.

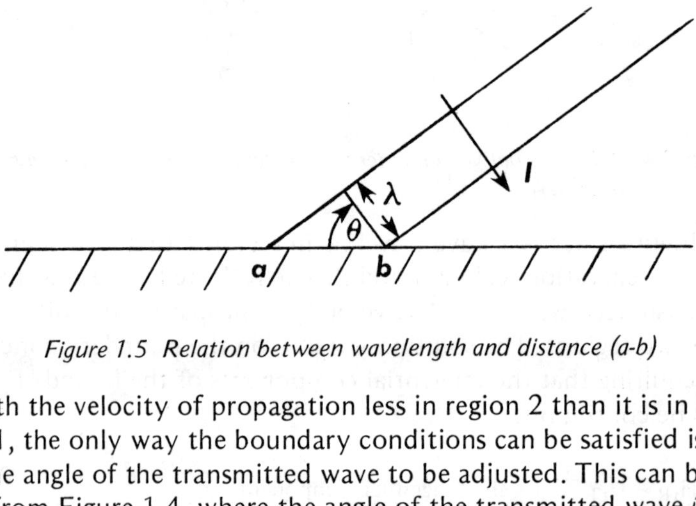

Figure 1.5 Relation between wavelength and distance (a-b)

With the velocity of propagation less in region 2 than it is in region 1, the only way the boundary conditions can be satisfied is for the angle of the transmitted wave to be adjusted. This can be seen from Figure 1.4, where the angle of the transmitted wave θ_T is greater than the angle of incidence θ_I. This is because the wavelength in region 2 is shortened conforming to the requirements that the velocity be less and the frequency be the same as compared to region 1. The distance (a-b) is common to the three waves. From Figure 1.5, $\cos\theta = \lambda/(a-b)$ where λ = wavelength.

$$\text{Then distance } (a-b) = \frac{\lambda_I}{\cos\theta_I} = \frac{\lambda_R}{\cos\theta_R} = \frac{\lambda_T}{\cos\theta_T} \tag{1.13}$$

The incident and reflected waves travel in the same medium so that $\lambda_I = \lambda_R$ and $\theta_I = \theta_R$, or the angle of reflection equals the angle of incidence. The angle of the transmitted wave can be found from Equation (6):

$$\frac{\cos\theta_I}{\cos\theta_T} = \frac{\lambda_I}{\lambda_T} = \frac{v_I}{v_T} = \frac{\mu_T\epsilon_T}{\mu_I\epsilon_I} \quad . \tag{1.14}$$

Relations among the waves at the boundary are shown in Figure 1.6.

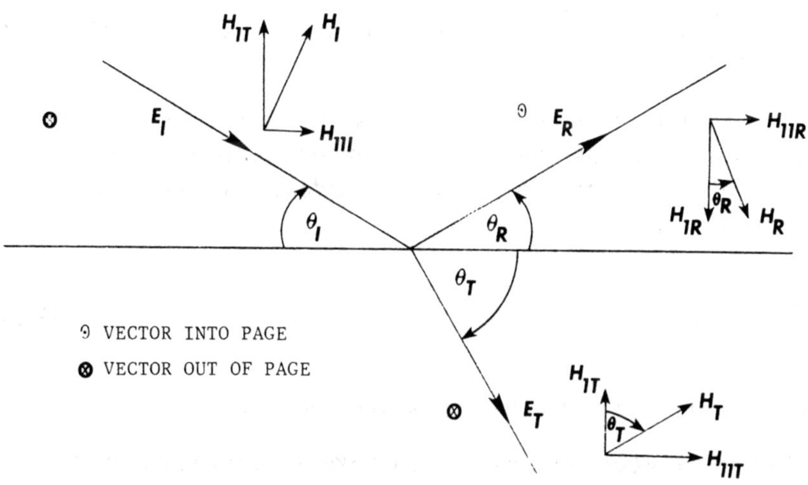

Figure 1.6 Relations between E & H fields at interface

For all plane waves, E and H are related through the intrinsic impedance n:

$$E_I = n_I H_I,\ E_R = n_I H_R,\ E_T = n_T H_T. \tag{1.15}$$

From these and Figure 1.6 we get:

$$H_{tI} = \frac{E_I}{n_I}\sin\theta_I,\ H_{tR} = -\frac{E_R}{n_I}\sin\theta_I,\ H_{tT} = \frac{E_T}{n_T}\sin\theta_T \tag{1.16}$$

The electric field is reflected with a reversal of sign so that the electric field at the surface is always zero. However, the magnetic field is reflected with unchanged sign so H_{tR} is negative when E_R is positive.

Substituting the above into the boundary equation for H gives:

$$\frac{E_I}{n_I} \sin \theta_I - \frac{E_R}{n_I} \sin \theta_I = \frac{E_T}{n_T} \sin \theta_T \tag{1.17}$$

Using the above with $E_I + E_R = E_T$ and some algebraic manipulations gives the relative magnitude of the reflected field as

$$\frac{E_R}{E_I} = \frac{n_T \sin \theta_I - n_I \sin \theta_T}{n_T \sin \theta_I + n_I \sin \theta_T} \quad \text{(for parallel polarization)} \tag{1.18}$$

The magnitude of the reflected wave is a function of the incident and transmitted wave angles, and the intrinsic impedances of the two mediums. For the case when the incident wave is normal to the plane of reflection:

$$\frac{E_R}{E_I} = \frac{Z - Z_o}{Z + Z_o} \tag{1.19}$$

where Z_o and Z are the intrinsic impedances n_I and n_T.

For an absorber, it is desirable to make the impedance of the material as close as possible to "free space" impedance. In this manner, the amount of energy reflected by the absorber is minimized.

In the above discussions, the E-field has been given greater attention than the H-field. This is because, for most substances that would be used for an absorber, more losses are developed through conductive heating than through magnetic displacement. Therefore, rather than magnetic permeability, the dielectric constant and reflection of the E-field are the important considerations.

The ideal absorber is a lossy dielectric whose dielectric constant is unity at the surface and gradually increases to a very high value with depth. The increase in dielectric constant is provided by a gradual increase in conductivity (for example as provided by carbon black) which attenuates the wave. The impedance then changes gradually so that no "impedance step" or discontinuity is seen by the wave and no reflection will take place. Because this ideal graded dielectric absorber is very difficult to manufacture, various geometrical shapes are employed to provide an equivalently smooth change in impedance. The most efficient of these is a pyramidally shaped absorber. Here the sharp point and gradual slope of the sides provide the desired transition through an ever increasing lossy medium. The back of the absorber should be a reflective surface, so that no energy can pass through. The depth of the absorber should be such that this reflective surface is effectively "hidden" from the wave. In other words, a high percentage of the wave should be absorbed by the time it arrives at the reflective surface.

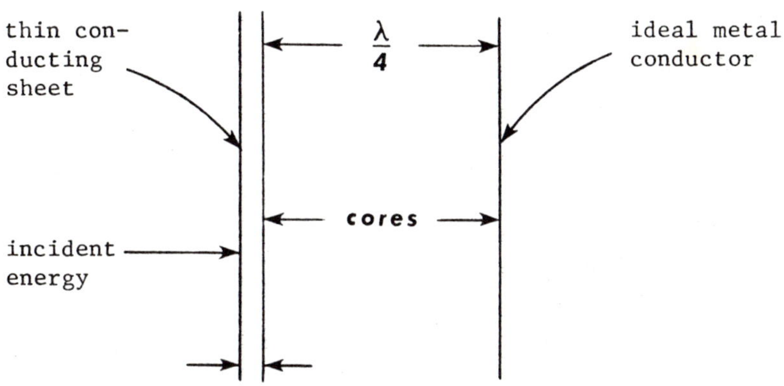

Figure 1.7 Salisbury Screen Absorber

Resonant Absorbers

The resonant absorber is variously known as the sandwich or Salisbury Screen absorber (see Figure 1.7). The principle is cancellation of the incident wave at the front face of the absorber.

The multicore sandwich absorber (based on quarter wavelength absorber) consists of the following:

a) An electrically thin front layer of such a conductivity and thickness that the energy reflected from it is negligible;

b) A highly conductive backing to ensure total reflection of electromagnetic waves;
c) An intermediate core filled with dielectric, the characteristics of which assume total phase and amplitude cancellation of the incident and reflected energy in the front layer.

$$t = \frac{\lambda_m}{4} \quad \frac{1}{4f\sqrt{\mu^1}} \tag{1.20}$$

t = thickness of λ/4 absorber in meters

f = frequency in cps

λ_m = λ of material in meters

A λ/4 absorber with a fixed thickness will absorb in narrow bands of f, 3f, 5f . . . and at odd harmonics of frequency.

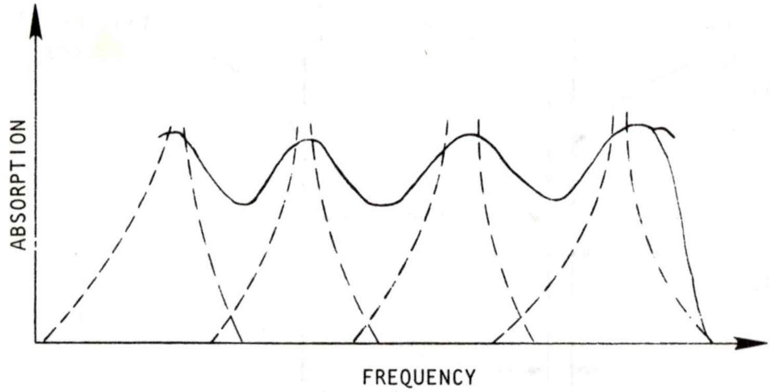

Figure 1.8 Multi-Layer Absorption Effect

The above describes the principle of sandwich absorbers, where each λ/4 absorbs most of the e.m. energy of a given frequency and adds to the absorption of other frequencies. To achieve wide bandwidth, a several multilayered system is used, as shown in Figure 1.8.

It should be noted, however, that the effectiveness of the absorber depends on the pattern of the illuminating antenna, and it is possible that one absorber might be better than another for a particular illumination.

NOMOGRAPHS AND TRADE-OFF CURVES

Absorbers come in many different shapes and sizes as required by their applications. Included in this section are curves representative of the various kinds of absorbers available on the market. In addition, there are nomographs giving the attenuation of lossy dielectrics and lists of commercial absorbers specifying the maximum power reflected (minimum power absorbed) for a given frequency.

Most specifications on absorbers give the guaranteed minimum absorption at a low frequency limit. Since absorption is dependent on frequency and the thickness of the absorber, absorption increases with frequency for a given absorber. Therefore, in general, an absorber will be more effective when used at higher frequencies than the low frequency limit specified.

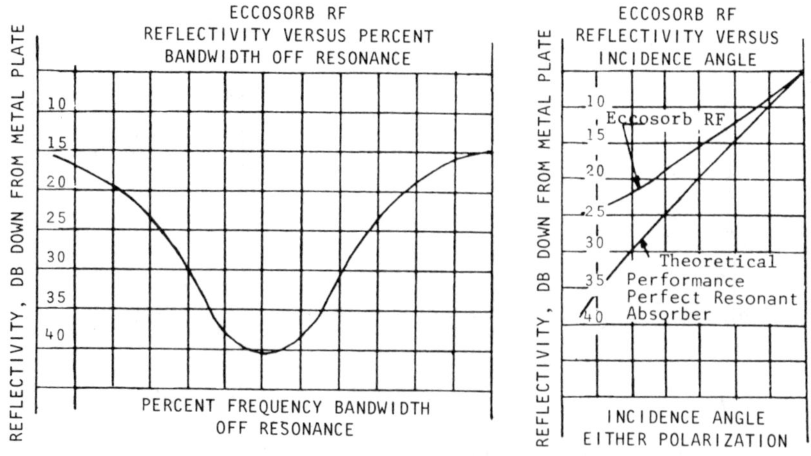

Figure 1.9 Absorption of resonant foam absorbers [1]

The tuned absorber or Salisbury Screen absorber, however, exhibits optimum performance only at the design frequency. An example of this type of absorber performance (for resonant foam) is shown in Figure 1.9.

Eccosorb RF is available tuned to any frequency in the S through Ku band region and exhibits absorption in excess of 25 dB at resonance bandwidth to the 5 percent power points; this bandwidth is about 35 percent and is adequate to cover a complete radar band. It can also be obtained for off-normal angles of incidence and for either polarization. Type RF is among those listed under "Commercially Available Absorber Materials".

The absorbers that exhibit the highest performance are the pyramidal types that have a sharp point, are high in depth relative to wavelength, and are made of a lossy foam material. Some have absorptivities that increase by approximately one order of magnitude for each octave of frequency. This performance extends at least into the 50-60 dB region. One type of pyramidal absorber is the Eccosorb VHP type and an example of its performance is shown in Figure 1.10.

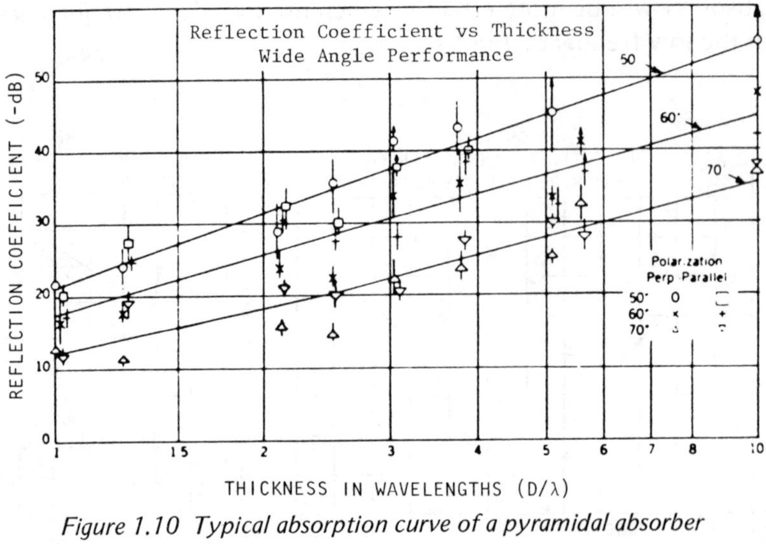

Figure 1.10 Typical absorption curve of a pyramidal absorber

Following is a list of available absorbers including a brief description of each. It is not meant to be a comprehensive listing but rather one that is representative of what is available.

For the case of a lossy dielectric used as an absorber, the nomograph in Figure 1.11 can be used to determine attenuation. The first nomograph can be used for dielectrics with loss tangents between 0.0001 and 0.05 and the second one (Figure 1.11(b)) used for loss tangents between 0.05 and 50.

Figure 1.12 gives curves of transmission loss of a half-wavelength sheet as a function of the loss tangent of the sheet.

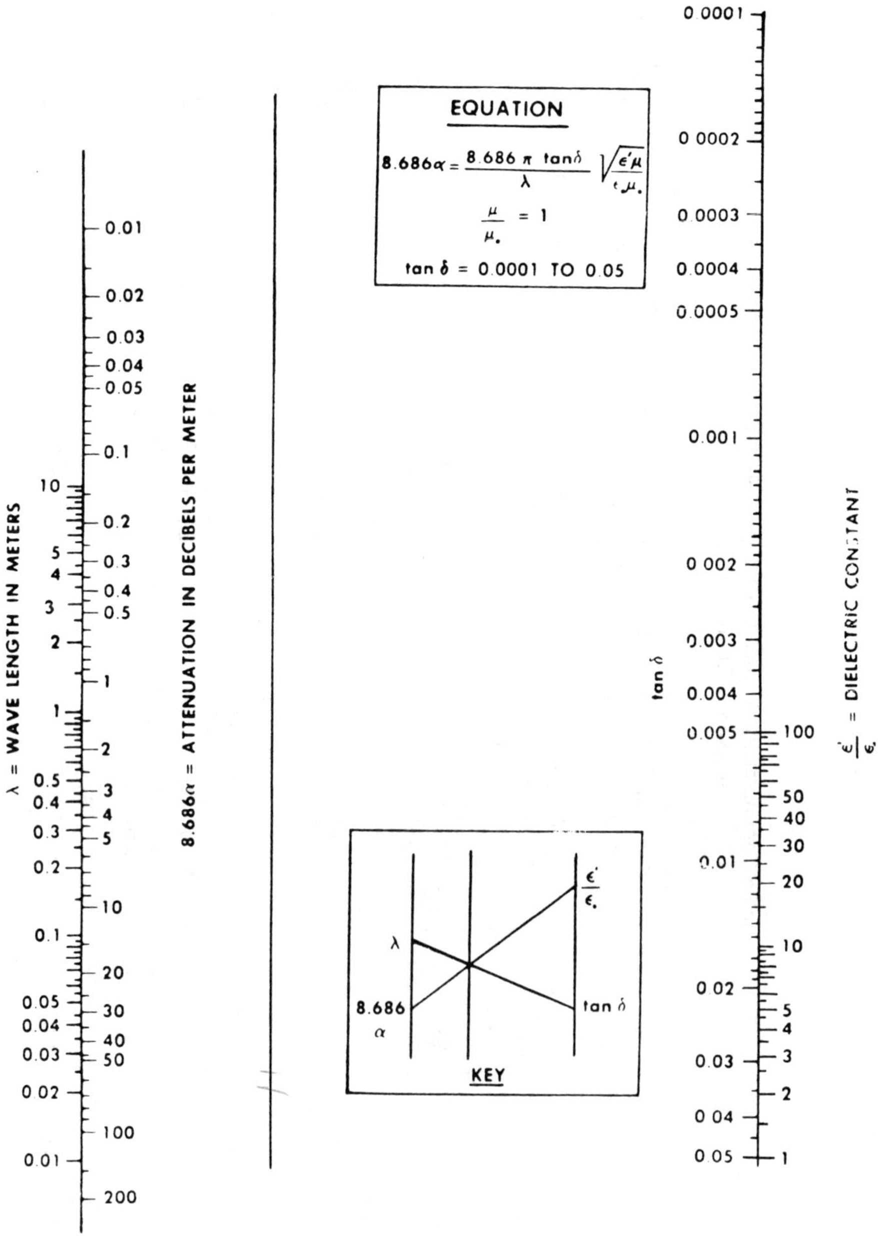

Figure 1.11A Attenuation in a dielectric with tan δ = 0.0001 to 0.5

TABLE 1.1

Some Commercially Available Absorber Materials
Emerson & Cuming, Inc.

Description	Type	Minimum Frequency or Frequency Range	Maximum Power Reflection	Thickness (Inches)	Weight (lb/ft^2)
ECCOSORB AN- Flexible foam sheer, light, easily cut	AN-72	20 GHz & up	−20 dB	1/4	0.1
	AN-73	7.5 GHz & up	−20 dB	3/8	0.2
	AN-74	3.5 GHz & up	−20 dB	3/4	0.3
	AN-75	2.4 GHz & up	−20 dB	1-1/8	0.5
	AN-77	1.2 GHz & up	−20 dB	2-1/4	0.9
	AN-79	0.6 GHz & up	−20 dB	4-1/2	2.0
ECCOSORB CV- Premium quality, electrically tapered	CV-4	2.5 GHz/25 GHz	−20 dB/−50 dB	4	0.8
	CV-6	1.5 GHz/25 GHz	−20 dB/−50 dB	6	0.9
	CV-9	1 GHz/25 GHz	−20 dB/−50 dB	9	1.8
ECCOSORB HPY- Ultra high performance, flexible foam pyramids	HPY-12	0.5 GHz/10 GHz	−20 dB/−50 dB	12	1.0
	HPY-18	0.4 GHz/7 GHz	−20 dB/−50 dB	18	1.2
	HPY-24	0.3 GHz/6 GHz	−20 dB/−50 dB	24	1.3
	HPY-30	0.2 GHz/5 GHz	−20 dB/−50 dB	30	2.5
	HPY-54	0.12 GHz/4 GHz	−20 dB/−50 dB	54	3.2
	HPY-72	0.08 GHz/3 GHz	−20 dB/−50 dB	72	3.7
ECCOSORB FR- Rigid polystyrene foam	FR-330	2.3 GHz	−20 dB	2	0.6
	FR-340	0.94 GHz	−20 dB	4	1.2

Table 1.1 (Continued)

Description	Type	Minimum Frequency or Frequency Range	Maximum Power Reflection	Thickness (Inches)	Weight (lb/ft²)
ECCOSORB H Curled rubberized fibers	H-1	5 GHz	−20 dB	1	0.3
	H-2	2.5 GHz	−20 dB	2	0.6
	H-4	1.0 GHz	−20 dB	4	1.0
	H-8	0.5 GHz	−20 dB	8	1.7
	H-12	0.3 GHz	−20 dB	12	2.3
ECCOSORB RM- Open cell silicone rubber foam, flat sheet	RM-75	2.4 GHz	−20 dB	1	1.0
ECCOSORB FBC- Coated fabric, hollow pyramids	FBC-75	3.0 GHz	−17 dB	1	1.0
ECCOSORB HT- High power, 650°F closed cell ceramic foam	HT-98	2.7 GHz	−17 dB	2	.15
	HT-99	0.94 GHz	−17 dB	3	0.7
ECCOSORB FGM- Ferrites in silicone rubber, flat sheet	FGM-40	4-10 GHz	−10 dB avg.	.04	1.0
	GFM-125	2-12 GHz	−12 dB avg.	.125	2.5
ECCOSORB NZ- Broadband, fired ferrite	NZ-1	50 MHz-12 GHz	−15 dB avg.	1	8.5
	NZ-2	.2 GHz/1 GHz/5 GHz	−15 dB/−10 dB/−5 dB	0.25	4.5
	NZ-31	0.25 GHz	−30 dB	0.29	7.7
	NZ-41	0.4 GHz resonance	−35 dB	0.275	6.4
	NZ-51	0.8 GHz	−35 dB	0.20	5.1

Table 1.1 (Continued)
Some Commercially Available Absorber Materials
Emerson & Cuming, Inc.

Description	Type	Minimum Frequency or Frequency Range	Maximum Power Reflection	Thickness (Inches)	Weight (lb/ft^2)
TYPE RF-Reinforced Urethane Plastic foam, designed to cover metallic surfaces	RF-KU	16 GHz	−25 dB	0.11	0.2
	RF-X	9-375 GHz	−25 dB	0.17	0.3
	RF-C	5.5 GHz	−25 dB	0.32	0.6
	RF-S	3 GHz	−25 dB	0.75	1.5
TYPE VHP-Very high performance, lightweight aggregate of foam pyramids. For use in anechoic chambers	VHP-70	.12/.2/.3/.5 GHz/L-band	−30/−35/−40/−45/−50 dB	70	6.0
	VHP-45	.2/.3/.5 GHz/L-/S-band	−30/−35/−40/−45/−50 dB	45	3.5
	VHP-26	.3/.5 GHz/L-/S-band	−30/−35/−40/−50 dB	26	2.0
	VHP-18	.5 GHz/L-/S-/C-band	−30/−40/−45/−50 dB	18	1.5
	VHP-8	L-/S-/C-/X-band	−30/−40/−45/−50 dB	8	1.0
	VHP-4	S-/C-/X-/Ku-band	−30/−40/−45/−50 dB	4-1/2	0.5
	VHP-2	C-/X-/Ku-/K-band	−30/−40/−45/−50 dB	3	0.45
TYPE SF-Resonant very thin, flexible Loaded Silicone or Urethane	24 Types	1-16 GHz	−20 dB	.035 to .22	0.6 to 6.0
TYPE MX 410 Resonant, flexible resistive front, metal foil back	410	9.375 GHz	−20 dB	.025	.156

Table 1. 2

Some Commercially Available Absorber Materials
Advanced Absorber Products, Inc.

Description	Type	Minimum Frequency or Frequency Range	Maximum Power Reflection	Thickness (Inches)	Weight (lb/ft^2)
TYPE ML Multilayer, broad-band flat, thin, flexible, varying electrical density	AAP-ML-72	16.0 GHz	−20 dB	0.25	0.2
	AAP-ML-73	8.0 GHz		0.375	0.3
	AAP-ML-74	4.0 GHz		0.75	0.4
	AAP-ML-75	2.5 GHz		1.125	0.6
	AAP-ML-77	1.5 GHz		2.25	0.8
	AAP-ML-79	0.5 GHz		4.5	1.8
TYPE ML-P Flexible closed cell foam, thin lightweight	AAP-ML-P-73	8.0 GHz	−17 dB	0.375	0.16
	AAP-ML-P-74	4.0 GHz		0.75	0.28
	AAP-ML-P-75	2.5 GHz		1.125	0.5
TYPE HC Treated glass hexagonal open cell honeycomb coated with lossy layer	AAP-HC-( ) AAP-HC-( )T AAP-HC-( )P (Pyramidal) AAP-HC-( )W (Wedge)	9.5 GHz	−12, −16 or 20 dB	Cut to order 3/16 or 1/4 inch cells	

Table 1. 2(Continued)
Some Commercially Available Absorber Materials
Advanced Absorber Products, Inc.

Description	Type	Minimum Frequency or Frequency Range	Maximum Power Reflection	Thickness (Inches)	Weight (lb/ft^2)
TYPE MC Multi-Cote broad band Structural honeycomb sandwich	MC	2-18 GHz	−17 dB	0.97	1.9
TYPE SS Thin, flexible tuned absorber metal foil backed	AAP-SS (standard)	2-4 GHz		.5 to 1	0.25
	AAP-SS-WR (weather resistant)	4-8 GHz	−23 dB	.25 to 0.5	0.32
	AAP-SS-WP (very flexible, weatherproof)	8-12 GHz		.1875 to .25	0.14
TYPE SF Thin, flexible sheets of loaded silicone rubber, resonant at specific frequencies	AAP-SF-9.5 (example at 9.5 GHz)	9.5 GHz	−20 dB	.0625	1.0

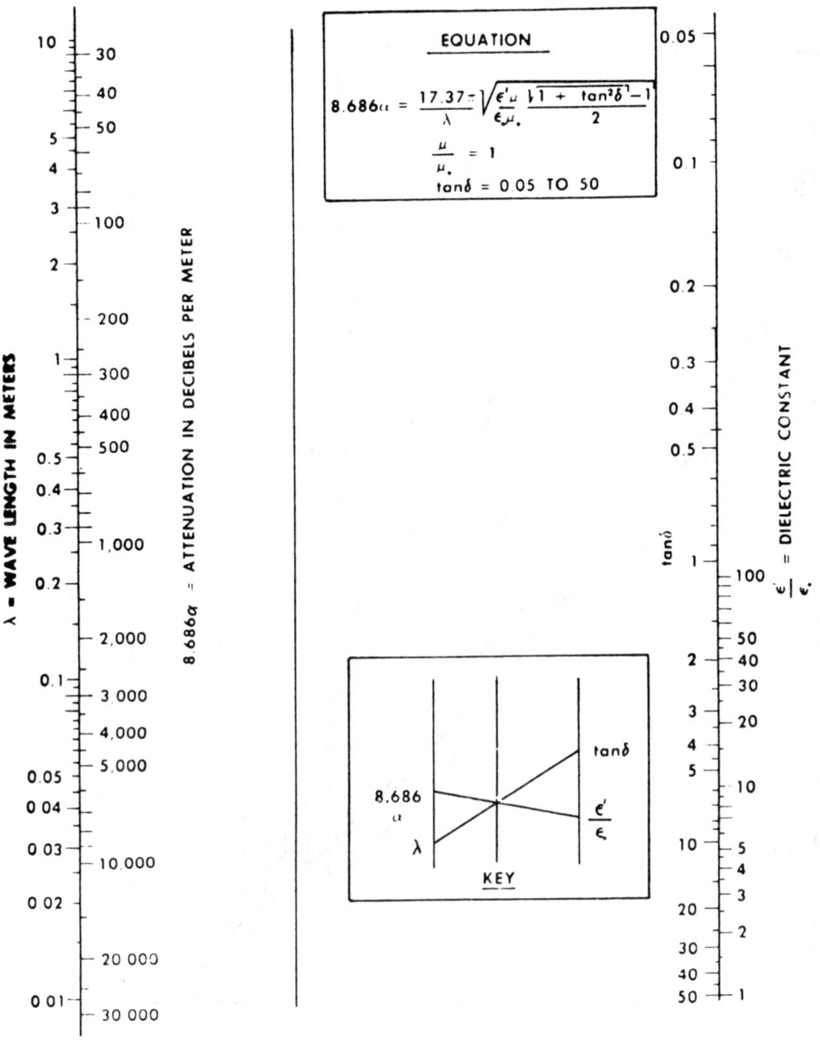

Figure 1.11B Attenuation in a dielectric with tan δ = 0.05 to 50

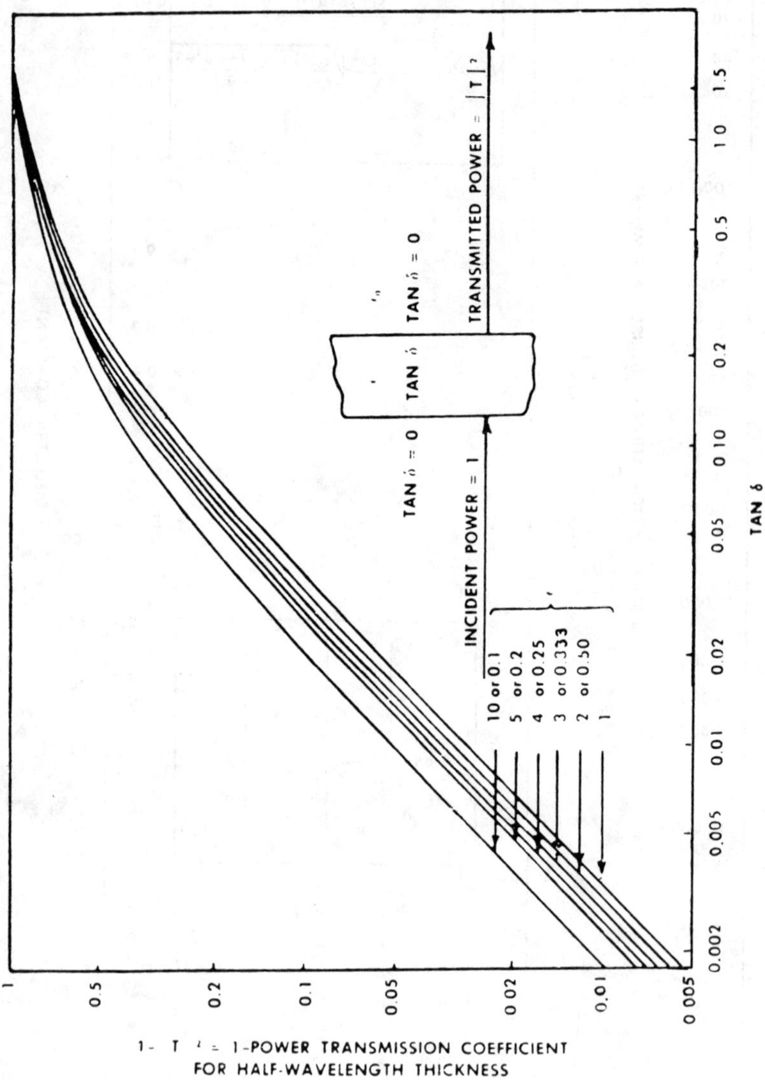

Figure 1.12 Transmission loss of a lossy sheet of half-wavelength thickness

Conclusions

For most applications, it would seem possible to provide at least −20 dB isolation of a radar from surrounding obstacles by placing RAM on the obstacles. When discrimination from a single frequency or radar band is all that is required, use of thin, lightweight tuned or resonant absorbers will work well. In thicker versions these types can be made broadband. For extreme interference problems between nearby radars, the use of fences and RAM lined tunnels or shields around Cassegrain dishes may give 40 to 50 dB isolation.

An additional way of using absorbing material is to place it along the edges of an antenna that are perpendicular to the E-field vector. When RAM was placed on the E-plane edges of a horn antenna the radiation was reduced 20 dB over the back 80° sector and 5 dB over the back 200° sector [2]. RAM was also applied to a parabolic antenna and the back lobe reduction was about 15 dB.

REFERENCES

[1] L. Peters, Jr., C. Rudduck, AD 600652 (Final Report) RADC TDR 63:549, "Application of Electromagnetic Absorbing Materials as Interference Reduction Techniques."

[2] L. Moreland and L. Peters, Jr., "The Specular Radar Cross-Section of Absorber Coated Bodies," *IEEE Trans.*, Vol. AP-14, Nov. 1966.

[3] L.J. Ladar, J.B. Winderman, "Method of Reducing Side Lobes of Directive Antennas," *Canadian Journal of Physics*, Vol. 44, 1966.

[4] J.A.M. Lyon, C.J. Digenis, W.W. Parker and M.A.H. Ibrahim, "Electromagnetic Coupling Reduction Techniques," University of Michigan, Radiation Laboratory, Ann Arbor, Michigan, June 1968.

[5] Emerson & Cuming, Inc., "Eccosorb Technical Information."

[6] Advanced Absorber Products, Inc., "Microwave Absorbers-Radomes-Special Products."

2

USE of RAM on PARABOLIC REFLECTORS

Radar absorbing materials may be used in reflector type antennas, such as the parabolic or Cassegrainian kinds, to provide additional illumination taper of the edge of the dish. The lower the illumination of the edge of the reflector, the lower will be the sidelobe levels. Work has been done [1] indicating that a modest reduction of the sidelobe levels is possible with the application of a limited amount of RAM to the edges of the reflector that are perpendicular to the E-field.

One reason that high sidelobes are generated is the occurrence of minor inaccuracies during fabrication which, because of cost considerations, are difficult to rectify. The inaccuracies will be expressed as quadratic phase errors to demonstrate the effects of amplitude tapering. Quadratic phase errors were chosen because they are the easiest, mathematically speaking, to treat. These errors actually occur when the feed of a parabolic antenna is not located at the focal or virtual focal point of the parabola. In general, the feed is not seen as a point source because of its finite physical size; the situation becomes worse as the diameter of the reflector, expressed in wavelengths, decreases.

For the analysis, an antenna with a circular aperture, a uniform amplitude distribution, and some phase error, β, of order n is assumed.

The gain function of such an antenna is [2]

$$g(\mu) = \int_0^1 r\, J_o(\mu r)\, e^{-j\beta r^n}\, dr \tag{2.1}$$

where the radius of the aperture has been normalized to unity. Also,

$$\mu = (\pi D/\lambda) \sin\theta \tag{2.2}$$

where D = diameter of the aperture

λ = wavelength

θ = angle off bore sight

r = distance along the radius

The one-way power gain is:

$$g(\mu) = g(\mu)\, g^*(\mu) \tag{2.3}$$

For $\beta \leqslant 1$, the phase error is quadratic and the original gain equation is now:

$$g(\mu) = \int_0^1 r\, J_o(\mu r)\, e^{j\beta r^2}\, dr \tag{2.4}$$

which, when solved gives:

$$g_p(\mu) = \left[\frac{J_1(\mu)}{\mu} - 4\beta^2 \frac{J_3(\mu)}{\mu^2}\right]^2 + \left[8\beta^3 \frac{J_4(\mu)}{\mu^4} - 2\beta \frac{J_2(\mu)}{\mu^2}\right]^2 . \tag{2.5}$$

RAM can be applied to the surface so that the loss increases as a function of radius. When this is done the amplitude distribution is tapered toward the edge of the reflector. The gain function now has to be modified to account for the non-uniform distribution and is now:

$$g(\mu) = \int_0^1 rF(r)\, J_o(\mu_r)\, e^{j\beta r^2}\, dr \tag{2.6}$$

where F(r) is the loss function, dependent on r.

The following three taper functions were investigated:

$$\text{Linear: } F(r) = \begin{cases} 0 & 0 \leqslant r \leqslant b \\ r\text{-}b & b \leqslant r \leqslant 1 \end{cases} \qquad b = \text{constant} < 1 \tag{2.7}$$

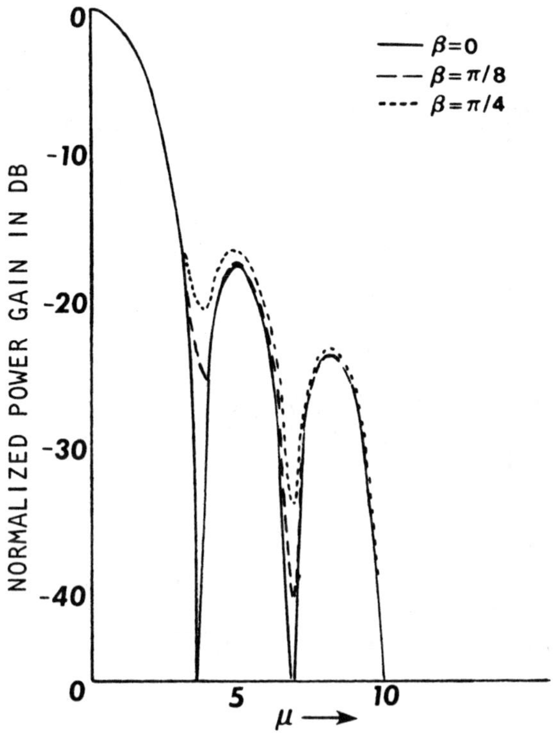

Figure 2.1 Calculated gain patterns of circular apertures with uniform amplitude distributions and quadratic phase errors of 0, π/8, and π/4 radians

Parabolic: $F(r) = 1 - r^2 \qquad 0 \leqslant r \leqslant 1$

Gaussian: $F(r) = \exp(-kr^2) \qquad 0 \leqslant r \leqslant 1$

the gain functions are now:

$$\text{Linear: } g(\mu) = \int_0^1 r\, J_o(\mu r)\, e^{j\beta r^2}\, dr - \int_b^1 r(r-b)\, J_o(\mu r)\, e^{j\beta r^2}\, dr \tag{2.8}$$

Direct solutions for the parabolic and Gaussian taper gain functions gives

$$\text{Parabolic: } g_p(\mu) = \left[2\frac{J_2(\mu)}{\mu^2} - 24\beta^2\frac{J_4(\mu)}{\mu^4}\right]^2 + \left[64\beta^2\frac{J_5(\mu)}{\mu^5} - 8\beta\frac{j_3(\mu)}{\mu^3}\right]^2 \tag{2.9}$$

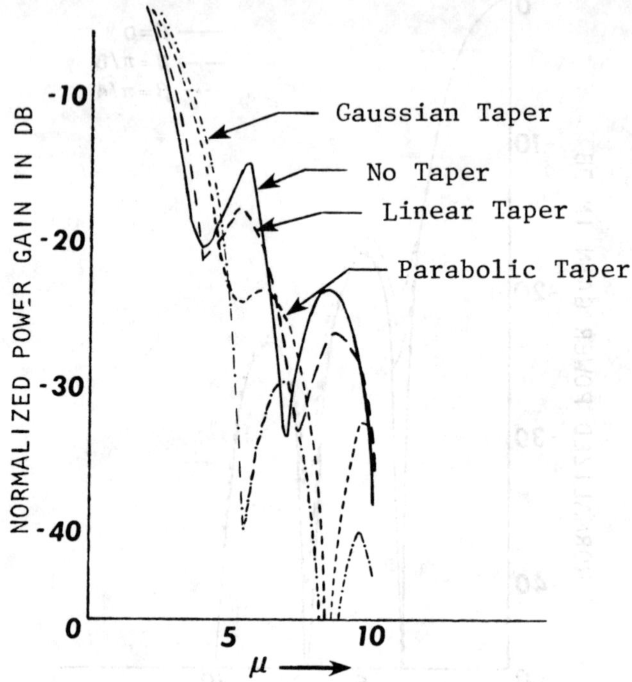

Figure 2.2 Calculated gain patterns of circular apertures with quadratic phase errors of π/4 radians, both treated and untreated

$$\text{Gaussian: } g_p(\mu) = e^{-2k} \left[\frac{J_1(\mu)}{\mu} + 2k\frac{J_2(\mu)}{\mu^2} + (4k^2 - 4\beta^2)\frac{J_3(\mu)}{\mu^3}\right]^2$$

$$+ \left[2\beta\frac{J_2(\mu)}{\mu^2} + 4\beta k\frac{J_3(\mu)}{\mu^3}\right]^2$$

The normalized antenna power gain patterns of three unmodified circular apertures with uniform amplitude distributions are given in Figure 2.1. One aperture pattern is ideal, the other two have quadratic phase errors of π/8 and π/4 radians respectively.

In Figure 2.2 the one-way power gain patterns of a circular aperture with a quadratic phase error of π/4 radians for the three taper functions is given.

The linear taper extended from r = 0.6 to r = 1.0. The Gaussian taper was of the form f(r) = exp $(-kr^2)$ and k was 4.6 so that the loss at the edge of the aperture was 20 dB greater than at the center. Some comparisons of the tapers give:

Type of taper	*Reduction in first sidelobes*	*Widening of main beam skirts at the -10 dB level (%)*
Linear	3 dB	2
Parabolic	9 dB	12
Gaussian	15 dB	14

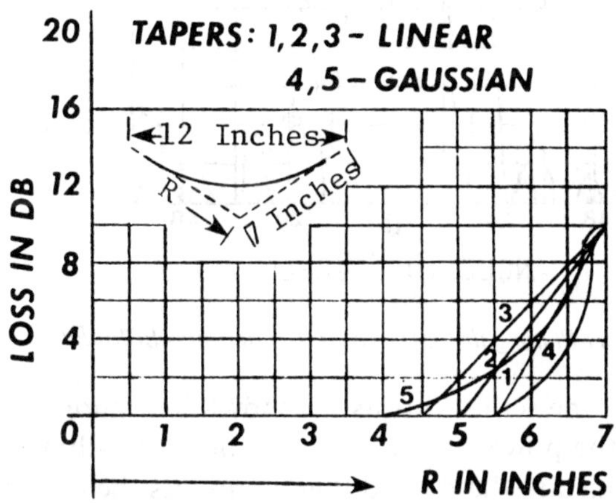

Figure 2.3 Contours of loss tapers

In no case was the loss of overall gain more than 1 dB. It is interesting to note that the taper that was most effective in reducing the sidelobes, the Gaussian taper, also had greatest effect on widening of the main beam. The energy that was not absorbed by the RAM near the edges would have been scattered in all directions, i.e., it would not have been properly focused into the main beam as it would have been had there been no absorbing material present. The result is that there is subsequent widening of the main beam and a slight loss of directivity.

Figure 2.3 shows several loss contours prepared for a 12″ paraboloidal antenna operated at X-band.

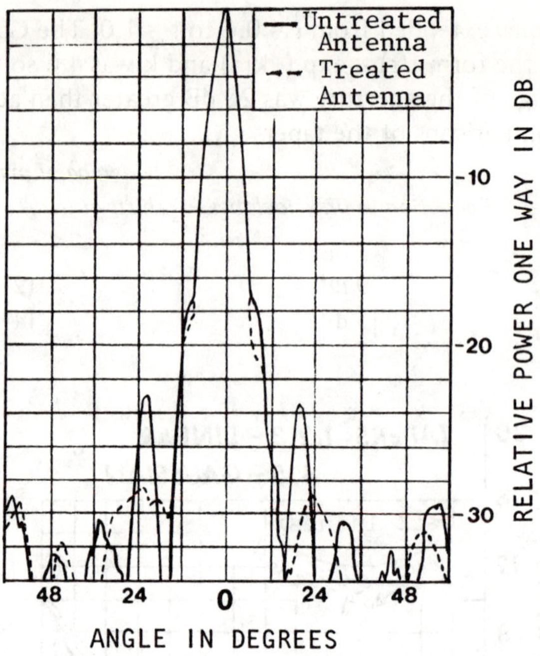

Figure 2.4 H-Plane gain patterns of 12-inch paraboloidal antenna

The results of one series of tests are shown in Figure 2.4, where the solid line is the power gain pattern of the optimally focused untreated antenna and the dashed line represents the pattern after application of one of the linear tapers. The sidelobes of the untreated antenna indicate that significant phase errors existed even after the antenna was carefully focused. These vestigial sidelobes were reduced by 1 and 3 dB each. The first major sidelobes were reduced by 5.5 dB at the expense of a 3/4 dB loss in gain. No widening of the main beam occurred. These results were obtained using one of the linear tapers and more sidelobe reduction should be possible by employing a Gaussian taper if some widening of the main beam can be tolerated.

Conclusions

Therefore, through a simple modification to those areas of the reflector that are perpendicular to the E-field, a sidelobe reduction to 5 to 10 dB, or even more, can be attained. The cost of such a modification should be reasonable and the additional weight on the reflector would be much less than a tunnel. The performance degradation that may result from adding radar absorbing material to the edge of the reflector is a slight widening of the 3 dB beamwidth.

REFERENCES

[1] L.J. Lader and J.B. Winderman, "Method of Reducing Side Lobes of Directive Antennas," *Canadian Journal of Physics*, Vol. 44, 1966.

[2] S. Silver, *Microwave Antenna Theory and Design*, New York: Dover, 1949.

[3] R.A.J. Ratnasiri, R.G. Kouyoumjian, and P.H. Pathak, "The Wide Angle Sidelobes of Reflector Antennas," Report 2183-1, 23 March 1970, The Ohio State University Electro Science Laboratory, Department of Electrical Engineering; prepared under contract AF 19(628)-5929 for Air Force Cambridge Research Laboratories (ATCRL-69-0413) (AD 707 105).

[4] C.E. Ryan and L. Peters, Jr., "Evaluation of Edge-Diffracted Fields Including Equivalent Currents for the Caustic Region," *IEEE Transactions on Antennas and Propagation*, Vol. AP-17, No. 3, May 1969, p. 292.

[5] Y. Hwang and R.G. Kouyoumjian, "A Dynamic Coefficient for an Electromagnetic Wave which is Rapidly-Varying at an Edge," URSI 1974 Annual Meeting, Boulder, Colorado.

[6] Y. Hwang and R.G. Kouyoumjian, "On the Dyadic Slope Diffraction Coefficient," URSI 1975 Annual Meeting, Urbana – Champaign, Urbana, Illinois.

Figures 2.2 and 2.3 Lader, L.J., and J.B. Winderman, "Method of Reducing Side-Lobes of Directive Antennas," *Canadian Journal of Physics*, Vol. 44, 1966.

3

FENCES

GENERAL DESCRIPTION

Introduction

The use of fences, i.e., reflective or absorptive barriers, has been a subject of interest for interference suppression for a number of years. The main concern in this regard is the development of a capability to determine the quantitative aspects of the change in coupling due to the insertion of a finite surface in the path between antennas. There are three major areas where such a capability would be important. They are: instrumentation, interference prediction, and interference suppression.

The use of such fences in instrumentation has been most prevalent in attempts to measure antenna patterns or signals where the goal is to evaluate only certain paths of interest, e.g., the path between a reflecting object and a measurement antenna. An example of such use is found in the measurement of three-dimensional antenna patterns by the sphere technique. This will not be described in detail in this report. In summary, the technique employs a balloon-supported reflective sphere, together with a measurement receiver and antenna designed to receive the reflected signal only, not the direct path signal from the transmitter. Double wire mesh fences were used to suppress the direct signal from the transmitter and suppression on the order of 60 dB was achieved at L-band frequencies [1].

The interest in the area of interference prediction is generated by the requirement to be able to account for the effects of fence-like barriers that would be found in the environment, such as aboard a ship or on a radar site. There is little if any theoretical work specifically addressed to modelling this situation.

The use of fences for interference suppression has been implemented for the purpose of suppressing self-interference, i.e., provide clutter rejection. The analysis of this specific problem has been, to a large degree, limited to the situation of very large aperture high gain land-based antennas where the high expense of a clutter fence could be justified. One such analysis is the study carried out by Wheeler Laboratories to determine a design for an anti-clutter fence for the AMRAD installation at White Sands Missile Range [3]. While the purpose of the fence is for the suppression of signal return from the surrounding mountains, the results can be interpreted in terms of coupling suppression. The AMRAD installation operates in the L-band.

The effects studied included the signals transmitted over, through, and under the fence; the effects of edge serrations, and the effect of the fence-reflected signal in regard to the tracking function. An empirical study based on a 93 GHz model was performed and the results were presented.

Radiation control fences have been used extensively lately due partially to the need for the acquisition of small targets at angles very close to the horizon. In this case, fences are employed to intercept ground clutter returns more efficiently than could be done by range-gating or filtering techniques such as MTI and doppler. The use of fences for RFI control is based on the interposition of an electrically opaque fence between an interfering source and the system to be protected. One example of such use could be found in the previously discussed protection of instrumentation antennas in measuring antenna patterns by the sphere technique [4]. Fences are also used to prevent reflected signals from reaching antennas and to protect personnel from radiation hazards.

A most interesting use of fences would be aboard ships to reduce interference from other nearby radars. Represented below is a configuration where one radar is above the other and where each represents an interference source to the other. The interposition of a radial fence, perhaps in the form of a platform, between the two radars will at least partially shield one from the other (see Figure 3.1).

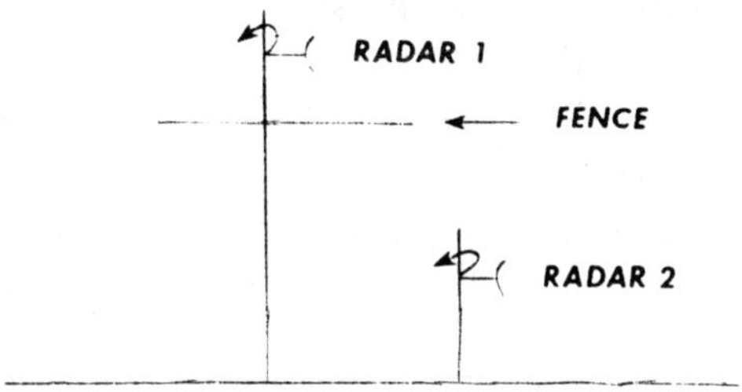

Figure 3.0 Fence-Like Barriers

Since radars scan an area of space, it is probable that, on occasion, they will point in the general direction of one another, thereby creating a mutual interference situation. In cases where there may not be a direct path of radiation, the spill-over or diffraction radiation around the reflector may be sufficient to cause problems because of the close proximity. The fence will stop nearly all of the direct interference through the fence but will diffract those rays near the edge. The ratio of radiation intensity with the fence to that without the fence will be derived below. This can be viewed as the attenuation introduced by the fence; numerically, the attenuation will be expressed in decibels.

Work has been done at the Georgia Institute of Technology [5] concerning scanning radars investigating the effects of objects placed in the Fresnel and Fraunhofer zones. These tests indicate that the effects of an object on the median gain are about the same in both the Fresnel and Fraunhofer zones. A few representative tables from this study are presented later in the chapter.

From this work we will assume, for a rotating antenna, that the effects of a fence placed in the near-field can be represented by far-field approximations.

When the antennas are nearly stationary, as they might be in tracking, and the fence is in the near-field, the incident field must be integrated over the antenna aperture to obtain the received signal. Comparison of the integrated fields with and without the fence gives the one-way clutter suppression. The latter integration gives the sidelobe level in the direction of the clutter or interfering source. If necessary a model can be written to account for the fence in the near-field of the antenna.

Another development in the field of fences is the double-slot fence described by Becker and Millet [7]. The slots are below the

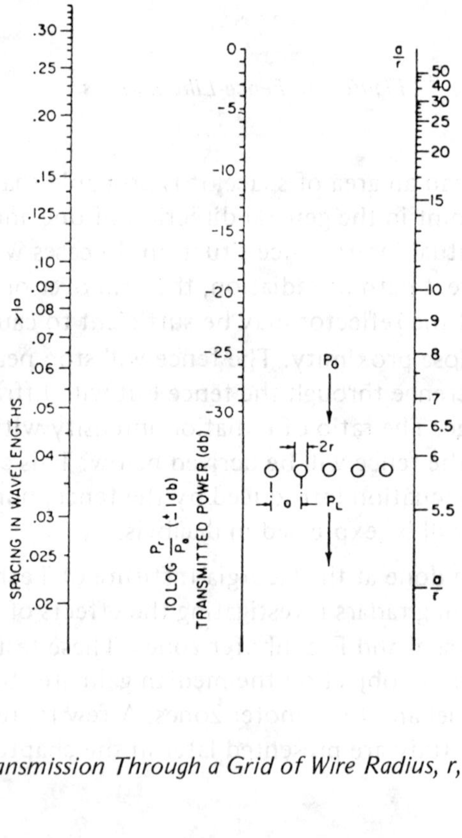

Figure 3.1 Transmission Through a Grid of Wire Radius, r, and Spacing, a

top edge of the fence and run the length of it. The object here is to make the equivalent radiation appear to come from a three-element line array. By proper adjustment of the slot dimensions and spacing of the slots, the total field from the line array can be made very small at the antenna aperture. However, this seems to be true for only a narrow range of clutter angles. This, then, would be useful against an interfering source or reflection whose position relative to the receiving antenna was fixed.

If the fence is made of solid metal there is no problem of leakage transmission through it. If, however, a fence is used large enough to be affected by wind conditions, then it must be of mesh-type construction. The attenuation of a mesh-type fence can be found from a nomograph given by W.W. Mumford [8]. If additional attenuation is required, a double-mesh fence will give considerable improvement over that of a single-mesh fence. If the attenuation of a single-mesh fence is α then, by transmission line analogy, a double-mesh fence will give an attenuation of $2\alpha + 6$ dB [9].

THEORY OF A STRAIGHT EDGE FENCE

Huygens-Fresnel Principle

From Huygens' construction, every point of a wavefront may be considered as the center of a secondary wavelet, with the wavefront at any later instant being the envelope of the wavelets (see Figure 3.2a).

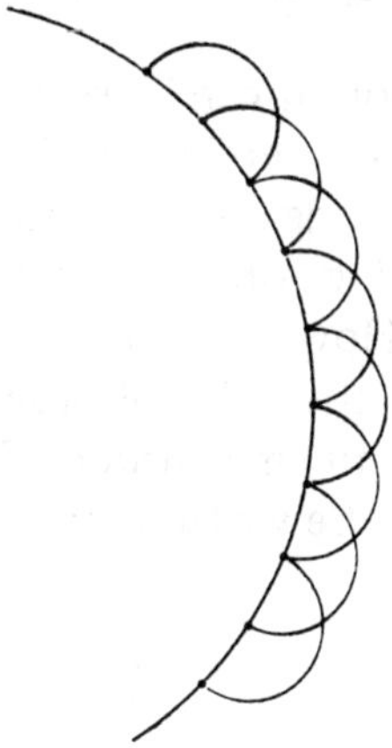

Figure 3.2a Huygens' Construction

Fresnel hypothesized that diffraction could be accounted for by considering that the secondary wavelets mutually interfere. This principle of interference together with Huygens' construction is known as the Huygens-Fresnel Principle.

In Figure 3.2b, A is the source of a spherical wavefront, B is the point of observation, and R is the distance from B to the wavefront. Fresnel constructed zones about B as the center of radius $R + \lambda/2$, $R + 2(\lambda/2)$. . . . In this manner zones were inscribed on the wavefront and are numbered 1, 2, 3, . . . in Figure 3.2b.

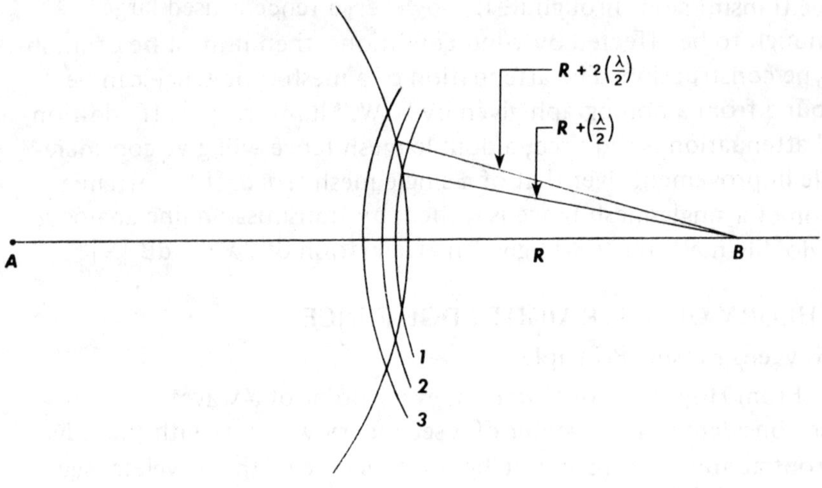

Figure 3.2b Fresnel's Zone Construction

From the construction, it can be seen that the total field at B can be approximated by summing up the contributions from each zone. Since each succeeding zone's contribution is 180° out of phase with the preceeding zone's contribution, it tends to cancel with the former zone. However, the succeeding zones are at increasing distances from the point of observation and, therefore, contribute less. Letting the magnitude of the nth zone be S_n, the total field at B can then be written as:

$$S = S_1 - S_2 + S_3 - S_4 + \ldots - S_n \tag{3.1}$$

Since S_n decreases with increasing n, the sum of two adjacent zones tends toward zero as n increases. It can be shown [13] that the total field at B is nearly equal to $S = 1/2(S_1 \pm S_n)$ depending on whether n is odd or even.

Zones are constructed until the line from B to the wavefront is tangential to the wavefront, as in Figure 3.3.

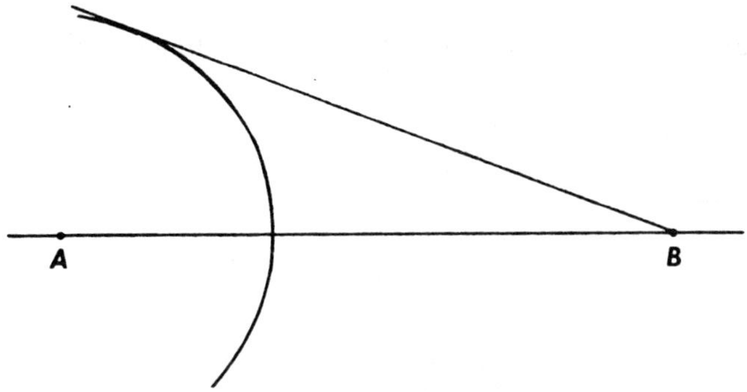

Figure 3.3 Limit of Zone Construction

When contributions from all zones that can be seen from B are added together, an interesting result occurs. For this condition, the field at B reduces to $S = 1/2(S_1)$, showing that the total field is equal to one-half of the field contributed by the first zone.

As R increases in value, additional zones are seen which contribute less; this results in less fluctuations of the field at B. The far-field or Fraunhofer region begins when the field at B is the result of the contributions of a fixed number of zones, i.e., the field at B is simply dependent on the distance from B to the wavefront. The number of zones seen from B is then a constant, and does not vary as R varies.

The previously discussed method of approximating diffraction as mutually interfering, secondary wavelets, results in an integral useful for calculating diffraction caused by obstacles. This integral is known as the Fresnel integral and it can be used to calculate the attenuation of a source signal by a straight edge or fence located between the source and the receiver. The fence effectively screens out the contributions from those wavelets whose direct path to the receiver is blocked by the fence.

For the general case of a distant interference source and a perfectly conducting half-plane, the theory of diffraction has been well documented in several articles [13] for both the rigorous solution and the Fresnel approximation. The difference in these solutions is a minor correction for polarization orientation.

The pertinent expressions for the "relative transmission" in the forward half-space are given below in terms of the Fresnel approximation.

The relative transmission in the presence of a fence is given by [2]:

$$\frac{E}{E_o} = \frac{\exp(j\frac{\pi}{4})}{2} F(v); \tag{3.2}$$

The Fresnel integral is defined by:

$$F(v) = \int_v^\infty \exp(-j\frac{\pi}{2}) \, d\tau \tag{3.3}$$

τ is a dummy variable of integration, and v is a variable dependent upon the problem geometry.

In this case v can be expressed by:

$$v = 2 \sin \phi/2 \quad \frac{2R_1 R_2}{\lambda(R_1 + R_2)} \tag{3.4}$$

where R_1 and R_2 are given as illustrated above, ϕ is the angle of diffraction and λ is the wavelength.

Referring to the Cornu Spiral, Figure 3.4, which has been reoriented to conform to the usual phasor conventions, the illumination in the various regions can be determined. The reference vector which corresponds to the direct signal in the absence of a fence, is drawn between the two spiral centers. The total field in the presence of a fence is measured from the tail of the reference vector to a point on the spiral corresponding to the appropriate value of v. In the shadow region (point A) the total field is just the diffracted field. In the illuminated region, the total field is the resultant of the direct and diffracted components.

For sufficiently large values of v, the Fresnel integral can be approximated by the following expression [6]:

$$F(|v|) \approx - \frac{-j \exp(-j \frac{\pi}{2} v^2)}{\pi |v|} \tag{3.5}$$

This approximation is plotted next to the Cornu Spiral.

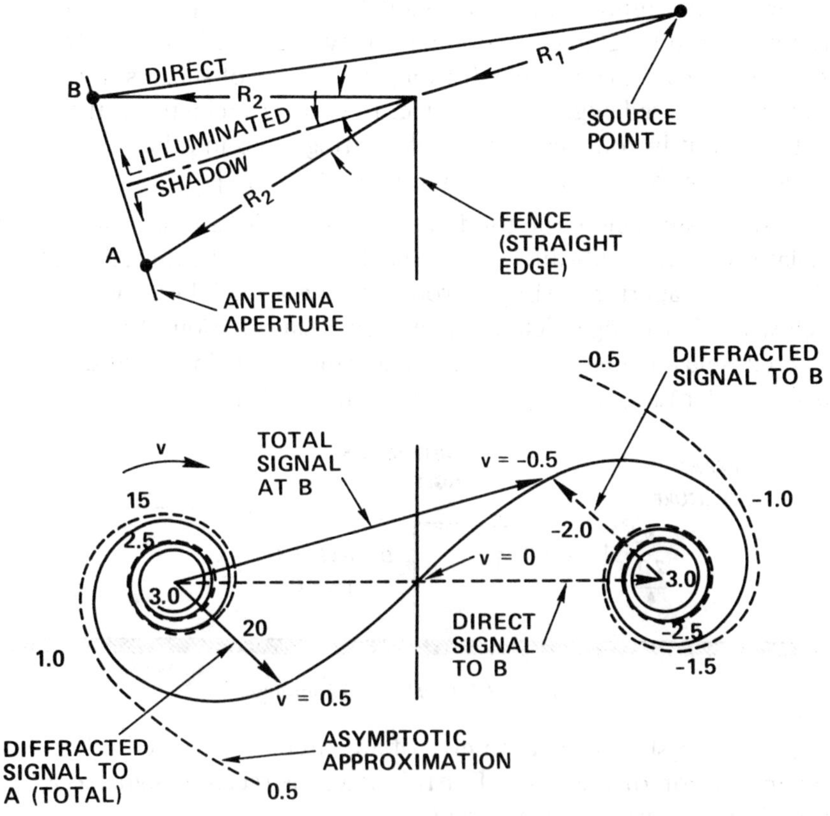

Figure 3.4 Graphical Representation of Diffraction by a Half-Plane

When the fence is in the far-field of the antenna (or equivalent to the same) the field incident over the antenna can be approximated by the field at the antenna center. However, when the fence is in the near-field, the received signal must be found by integrating the incident field over the antenna aperture.

$E(\theta) = \int\int f(x, y, \theta)\, v(x, y, \theta)\, dxdy$

$f(x, y, \theta)$ = transmitter antenna illumination

$v(x, y, \theta)$ = receiving antenna incident field

θ = antenna pointing angle (3.6)

Theory of a Double-Slot Fence

Significant improvement over straight edge fence performance can be gained by going to a double-slot type fence. Various edge treatments have been tried with varying degrees of success. For instance, a serrated edge fence reduces the power incident on the antenna, but introduces undesirable grating lobes in other azimuth directions, as has been shown by Lincoln Lab [14].

A single slot below the top fence edge provides a first-order cancellation of the fields diffracted over the fence in the direction of the antenna aperture. Also, it produces no grating lobes. Two slots below the fence edge yield a second-order cancellation of the diffracted fields incident on the antenna. This results in operation over a wider range of angles than with the single slot.

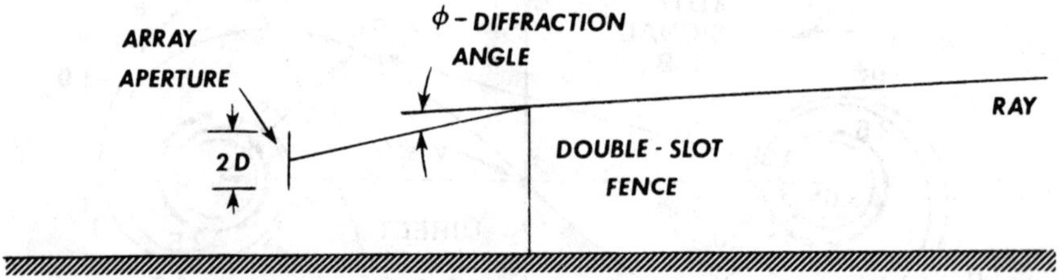

Figure 3.5 Pertinent Geometry

Figure 3.5 shows the geometry of an antenna aperture in the shadow region of the fence. For this case, the received signal as a function of elevation scan angle θ is

$$S(\theta) = \int_{-D}^{D} f(x;\theta)\, E(x)\, dx$$

$f(x;\theta)$ = the array aperture illumination function

$E(x)$ = incident field diffracted from the fence

x = aperture coordinate in the vertical plane. (3.7)

Starting with straight edge diffraction theory, the diffracted field relative to the field with no fence can be found from the Fresnel integral as:

$$E(x) = \exp(j \frac{\pi}{4}) \frac{F(v)}{\sqrt{2}}$$

$$F(v) = \int_{v}^{\infty} \exp(-j \frac{\pi}{2} \tau^2) \, d\tau$$

$$v = 2 \sin \frac{\phi}{2} \quad \frac{2R}{\lambda} \tag{3.8}$$

From array theory it can be shown that three line sources can be utilized to give a tapered field distribution across distant antenna aperture. Such an arrangement of sources with the resultant field distribution is given in Figure 3.6.

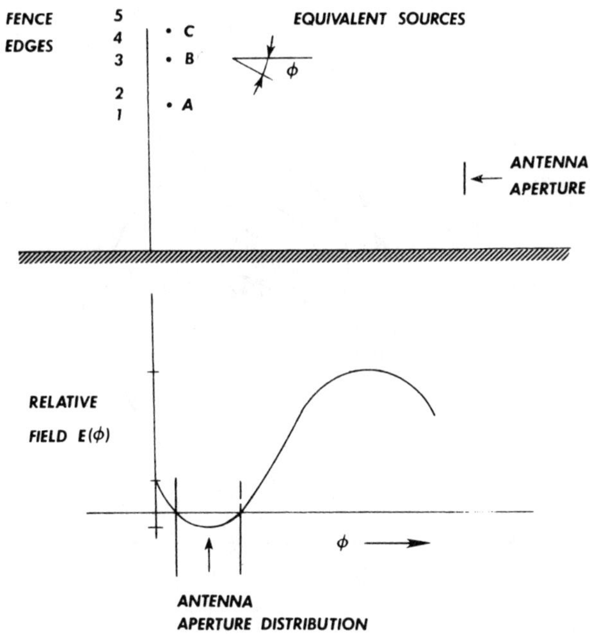

Figure 3.6 Field Distribution of Equivalent Line Source

When the fence is in the far-field of the antenna the diffracted field amplitude is approximately a constant over the aperture. This yields

$$S(\theta) \cong |E(x_o)| \int_{-D}^{D} f(x;\theta) \exp(j\angle E(x))\, dx \tag{3.9}$$

It can be shown that the phase of E(x) can be approximated as that resulting from a wave at the fence edge. The remainder of (1) within the integral is the aperture illumination function. Therefore, the response to the clutter return is the same as that for a free-space pattern. The peak response is in the direction of the fence edge, not the clutter source.

When |E(x)| is not nearly constant, the clutter suppression of the fence is determined not only by the average amplitude of the diffracted field across the aperture, but also by the distribution of the diffracted field.

Since it is desired to reduce the sidelobe clutter response, the field distribution should be tapered at the off-axis angles. Then the

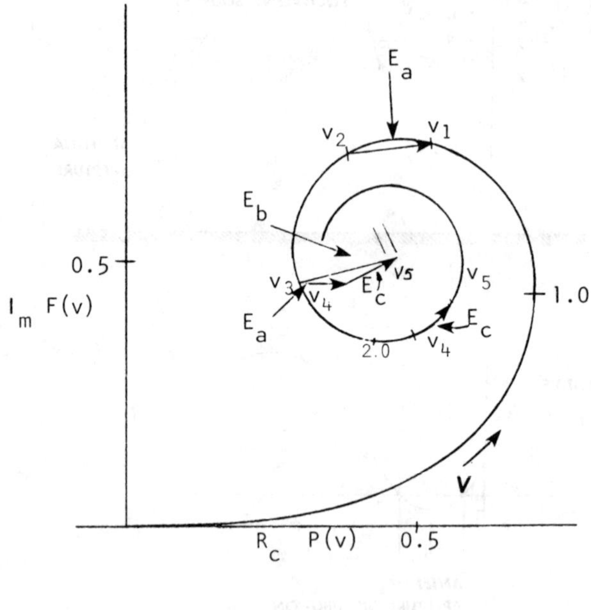

Figure 3.7 Graphical Summation of Equivalent Line Sources

product of the diffracted field distribution and the impressed antenna illumination is equivalent to a distribution which has a low sidelobe response. The pair of line radiators A and C have sinusoidal distributions while B is a constant. By adjusting the spacing between A and C, the minimum of the relative field will occur at the aperture center. By adjustment of the amplitudes of A and C relative to B the pattern will have zero amplitude at the aperture edges. The resulting field distribution has a lower average value across the aperture than for first-order cancellation and the shape of the distribution results in low sidelobe response.

By appropriate adjustment of the fence edges relative to each other, the vector addition of the three equivalent line sources can be made small over a given range of clutter angles. The radiation can be calculated from the Fresnel integral as follows:

$$\sqrt{2E_A} = F(v_1) - F(v_2)$$

$$\sqrt{2E_B} = F(v_3)$$

$$\sqrt{2}E_C = -[F(v_4) - F(v_5)] \tag{3.10}$$

The radiation from the equivalent line sources can be shown on the Cornu spiral, which is a graphical representation of the Fresnel integral. The vectors on the spiral corresponding to the diffraction from sources A, B, and C are shown in Figure 3.7. The phasors illustrated in Figure 3.7 pertain to the angle of diffraction corresponding to the center of the aperture. As the diffraction angle changes, the angles of the phasors vary in a manner shown in Figure 3.7.

TRADE-OFF CURVES AND RESULTS

Work has been done at the Georgia Institute of Technology [5] which indicates that for rotating antennas the median gain does not appreciably change in going from the far-field to the near-field of the antenna. However, the standard deviation of the gain increases with decreasing range to the antenna aperture.

Data is presented in tabular form to show the trend of the statistics as the antenna is probed from 50 feet to just short of physical contact. Far-field data is presented in most cases for comparison purposes. Statistical data as a function of range for the front 180° sector (or within ±90° from the peak of the main lobe), the back 180° sector and the total 360° pattern of the antennas are

Table 3.1 Statistical Characteristics of the S-Band Model Antenna as a Function of Range

RANGE $\frac{R}{2D^2/\lambda}$	STATISTICAL CHARACTERISTICS* Front 180°		Back 180°		Total 360°	
	m(dB)	*σ(dB)*	*m(dB)*	*σ(dB)*	*m(dB)*	*σ(dB)*
1.295	- 9.5	5.0	-13.5	4.5	-11.5	5.0
0.602	-10.7	6.0	-14.8	4.5	-13.0	5.0
0.217	-11.0	5.5	-14.9	4.5	-13.1	5.5
0.170	-12.3	6.0	-17.0	6.0	-15.1	6.5
0.118	-12.6	6.5	-17.5	4.0	-15.9	5.5
0.065	-12.0	5.0	-22.4	6.0	-16.8	8.0

*The median gain is designated as m and the standard deviation as σ.

Table 3.2 Statistical Characteristics of the GT/AR-4C Antenna as a Function of Range

RANGE $\frac{R}{2D^2/\lambda}$	STATISTICAL CHARACTERISTICS* Front 180°		Back 180°		Total 360°	
	m(dB)	*σ(dB)*	*m(dB)*	*σ(dB)*	*m(dB)*	*σ(dB)*
3.800	not recorded		not recorded		-11.1	8.0
0.269	- 4.4	6.0	-16.9	4.5	-10.0	8.5
0.185	- 4.7	8.5	-16.8	4.5	-12.8	8.5
0.128	- 5.0	7.0	-18.4	4.5	-12.5	9.5
0.055	- 6.5	6.0	-18.2	4.5	-13.8	9.5
0.020	- 5.8	6.0	-23.0	7.0	-13.2	11.5

*The median gain is designated as m and the standard deviation as σ.

presented. The antennas used were the S-band model, the GT/AR-4C, the GT/AR-6C, the x-band breadboard model and the GT/AR-4x.

Table 3.3 Statistical Characteristics of the GT/AR-6C Antenna as a Function of Range

RANGE $\frac{R}{2D^2/\lambda}$	STATISTICAL CHARACTERISTICS m(dB)	σ(dB)
1.695	-10.7	8.0
0.120	-10.5	9.0
0.083	-10.2	8.0
0.057	-10.8	7.5
0.036	-10.4	8.0
0.030	-10.8	8.0
0.021	-10.6	9.0
0.015	-10.1	10.0
0.009	-11.2	13.0

*The median gain is designated as m and the standard deviation as σ.

Table 3.4 Statistical Characteristics of the X-Band Breadboard Model Antenna as a Function of Range

RANGE	STATISTICAL CHARACTERISTICS*					
$\frac{R}{2D^2/\lambda}$	Front 180°		Back 180°		Total 360°	
	m(dB)	σ(dB)	m(dB)	σ(dB)	m(dB)	σ(dB)
1.495	-21.2	5.5	-25.9	5.5	-23.5	6.0
0.106	-18.3	6.0	-27.5	5.0	-23.2	8.0
0.051	-19.0	6.5	-29.2	5.5	-24.5	8.0
0.020	-21.0	6.0	-30.3	5.0	-25.5	8.0
0.013	-20.8	6.0	-31.5	5.0	-26.7	8.5
0.009	-19.6	8.0	-30.5	6.5	-26.0	9.5
0.006	-15.0	15.2**	-34.5	5.5	-28.0	17.0**

Table 3.5 Statistical Characteristics of the GT/AR-4X Antenna as a Function of Range

RANGE $\frac{R}{2D^2/\lambda}$	STATISTICAL CHARACTERISTICS		
	m(dB)	*$\overline{m}$(dB)*	*σ(dB)*
2.240	- 9.4	-10.0	11.4
0.159	- 9.8	-10.9	12.0
0.077	- 9.4	-10.6	12.4
0.028	-11.0	-12.4	12.4
0.019	- 9.7	-11.8	13.2
0.013	-13.0	-14.1	13.9
0.011	-12.7	-15.1	15.7

*The median gain is designated as m, the mean gain as $\overline{m}$, and the standard deviation as σ.

This indicates that it may be reasonable to assume that the attenuation of an interference signal by a fence can be determined by using far-field approximations, although the antennas are not in the far-field of each other.

Becker and Sureau have done some work with straight-edge fences in reducing ground clutter returns. Their work concerned large ground based radar where the source of interference was a distant hill and the fence was placed in the path of direct interference. Figure 3.8 represents a typical situation and the associated graph indicates the one-way clutter signal suppression as a function of normalized fence distance and height.

When the antenna-to-fence angle (ϕ) is a constant, or when the fence distance to fence height ratio is a constant, and the fence dimensions are increased, the amount of attenuation is increased. However, this also increases the cost of the fence. Since there is diffraction over the fence edge, there exists a pattern variation in the angular region just above the edge. Therefore it is desirable to minimize the angle subtended by the fence from the antenna. The result is that the amplitude of the diffracted signal decays more rapidly with the angle of diffraction for large antenna-to-fence distances than for small ones. For maximum suppression, it is desirable to use large fence dimensions, but cost and other considerations will dictate the upper bounds.

To check out the idea of a simple fence a computer program was written. The results compared favorably with those in Figure 3.8.

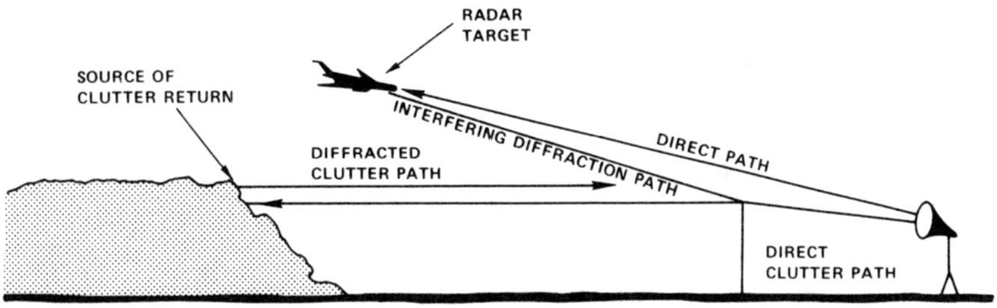

Figure 3.8 Clutter suppression (one-way) for straight-edge fence in the far-field of the antenna

The situation shown in Figure 3.9 was modeled.

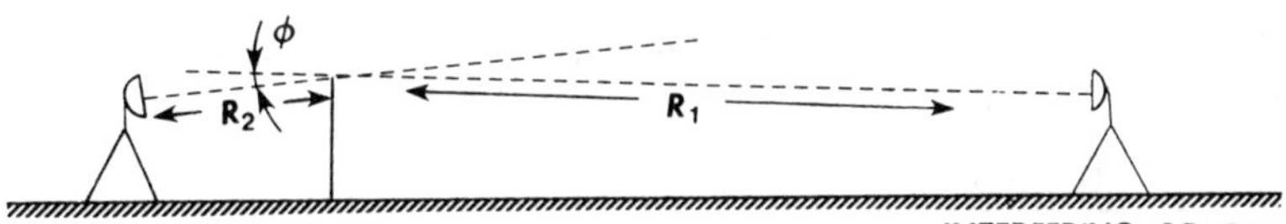

Figure 3.9/Problem geometry

Here ϕ is the angle of diffraction, R_1 the distance from the fence edge to the interference source, and R_2 the distance from the fence edge to the receiver. The ratio of signal levels with and without the fence is

$$\frac{E}{E_o} = \frac{\exp \frac{\pi}{4}}{2} F(v) \tag{3.11}$$

where F(v) is the Fresnel integral and v is a function of the geometry given by

$$v = 2 \sin \frac{\phi}{2} \quad \frac{2R_1R_2}{\lambda(R_1 + R_2)} \tag{3.12}$$

For sufficiently large values of v ($v \geqslant 2$) an approximation is valid for the Fresnel integral and is

$$F|v| \approx \frac{-j \exp(-j(\pi/2)v^2)}{\pi v} \tag{3.13}$$

We are interested in real power, so

$$\frac{E}{E_o}^{2} \approx \frac{F^2(v)}{4} \tag{3.14}$$

and the magnitude of

$$F^2(v) \approx \frac{1}{\pi^2 v^2} \tag{3.14a}$$

Substituting (3.12) and (3.14a) and (3.14) gives the power attenuation as

$$\frac{P}{P_o} = \frac{\lambda(R_1 + R_2)}{8\pi^2 R_1 R_2 (1 - \cos\phi)} \tag{3.15}$$

Requiring v to be greater than or equal to 2 is the same as requiring

$$\frac{R_1 R_2 (1 - \cos\phi)}{R_1 + R_2} \geqslant \lambda \tag{3.16}$$

Equation (3.15) gives the attenuation of the undesired signal due to the fence and will be expressed in decibels. Figure 3.10 shows curves of one-way interference suppression for various antenna-to-fence distances as a function of fence height. Fence dimensions are expressed in terms of wavelengths. Fences are frequency dependent in that they provide greater attenuation for higher frequencies.

Figure 3.11 expresses the relationship between fence distance (from antenna) and fence height for constant attenuation. This curve compares favorably with the result presented by Becker and Sureau (referenced earlier) for a similar situation that was modeled. This was done to check the model.

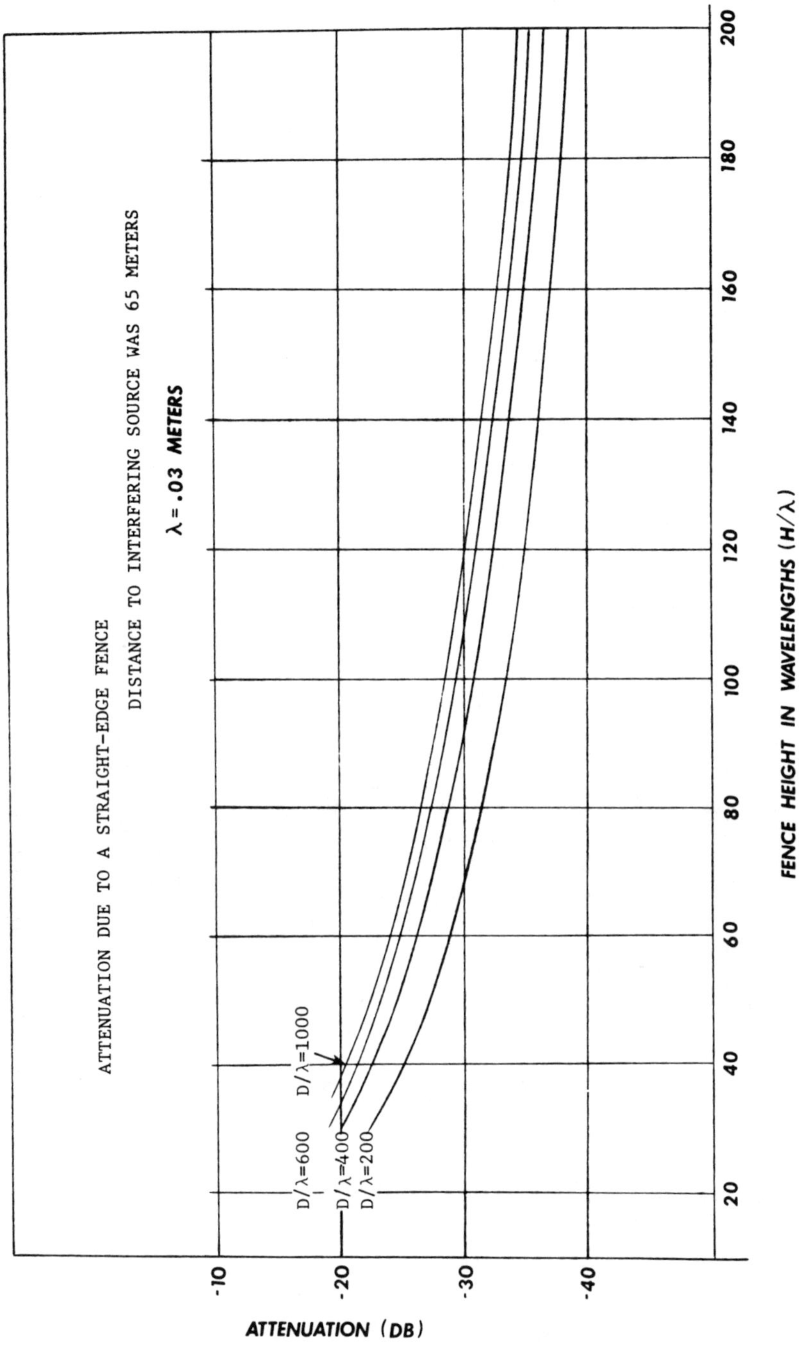

Figure 3.10/Attenuation of a straight-edge fence

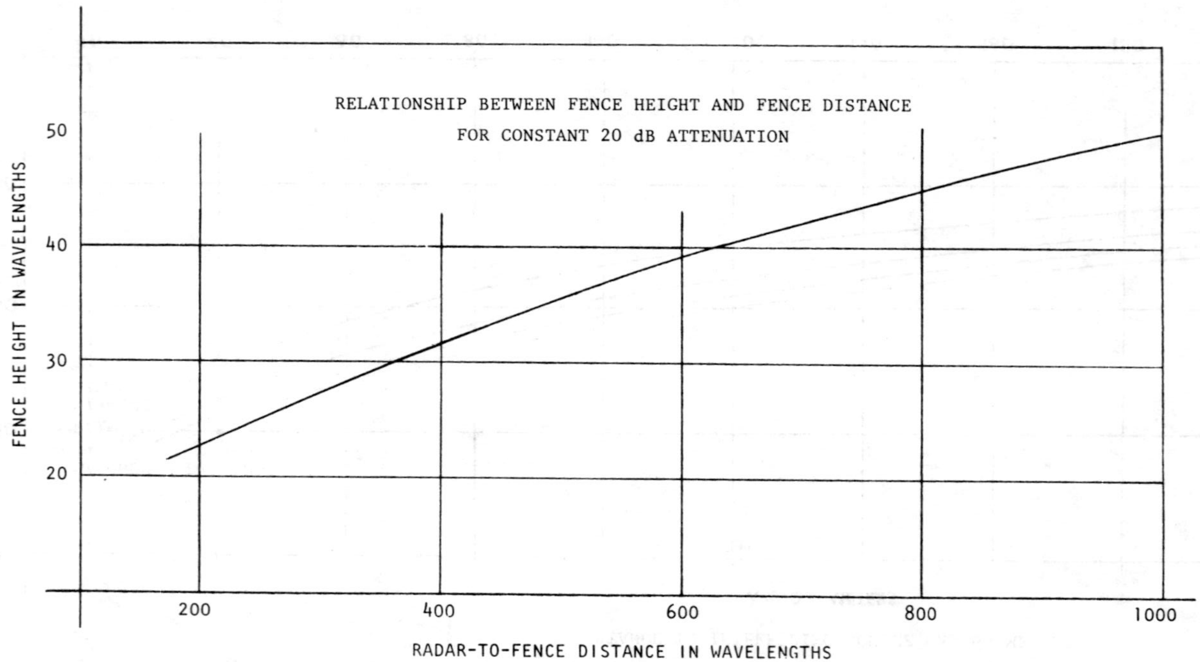

Figure 3.11/Relationship between fence height and fence distance for constant attenuation

The results presented here may be interpreted as the sort of results that are possible in the use of fences. No particular shipboard situation or antenna was considered although Figure 3.10 would be more realistic as far as the interference source distance is concerned.

The Double-Slot Fence

Several fence edge treatments have been considered in the literature, such as putting strips above the fence edge [16], or simple serrations [17] below the fence top (see Figure 3.12).

In a double-slot fence the idea is to make the slots appear as three line radiators whose average field is minimized over the antenna aperture and tapered at its edges. Tapering the field at the antenna edges reduces sidelobe response (see Figure 3.13).

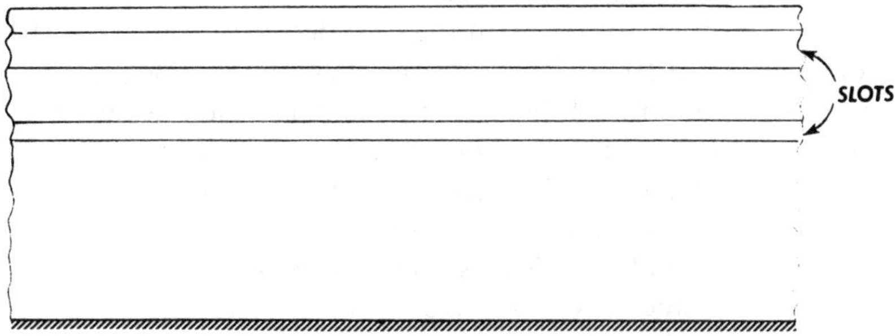

Figure 3.12/Double-slot fence

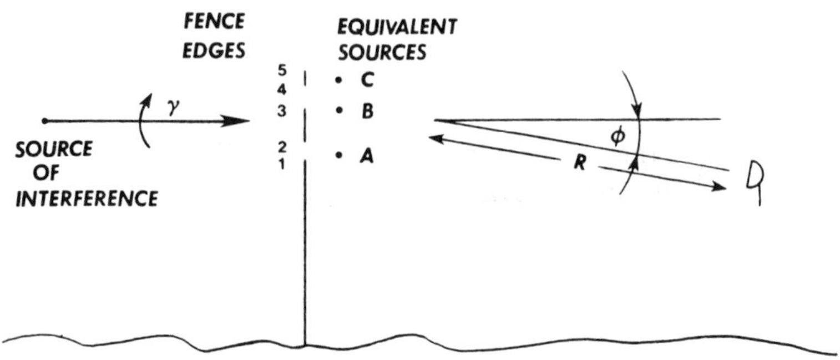

Figure 3.13/Equivalent line sources for a double slot fence

For a straight-edge fence, the diffracted signal at a point in space relative to the signal with no fence is given in terms of the Fresnel integral as

$$\mathrm{EXP}\ -j\frac{\pi}{4} = \frac{F(v)}{\sqrt{2}} = \frac{1}{\sqrt{2}}\int_{v}^{\infty} \mathrm{EXP}\ -j\frac{\pi}{2}\tau^2\ d\tau \qquad (3.17)$$

where, for this case,

$$v = 2\sin\frac{\phi}{2}\ \frac{2R}{\lambda} \qquad (3.18)$$

For large values of v, the diffraction from a straight-edged obstacle can be approximated by radiation from a line source. Thus, the radiation from slots beneath the top edge of the fence is nearly equivalent to that from a multi-element line source. For the fence shown certain additional simplifications can be made. When edge 1 is close to edge 2 the two line sources are equivalent to a source at A. The same can be done for edges 4 and 5; they are replaced by a source at C. The source at edge 3 is represented by a source at B. The interference radiation is now seen as coming from the equivalent sources located at A, B and C and their radiation can be determined from the Fresnel integral as follows:

$$\begin{aligned} \sqrt{2}\,E_A &= F(v_1) - F(v_2) \\ \sqrt{2}\,E_B &= F(v_3) \\ \sqrt{2}\,E_C &= -[F(v_4) - F(v_5)] \end{aligned} \qquad (3.19)$$

The total field is the vector summation of the individual fields, or in terms of the Fresnel integral

$$\frac{E_{Total}}{E_o} = \frac{E_A + E_B + E_C}{E_o}$$

$$= \frac{\text{EXP}\ j\frac{\pi}{4}}{\sqrt{2}} \left[F(v_1) + F(v_3) + F(v_5) - F(v_2) - F(v_4)\right)$$

$$= \frac{\text{EXP}\ j\frac{\pi}{4}}{\sqrt{2}} F_{Total}(v) \tag{3.20}$$

Let v_i be the ith source, where $1 \leqslant i \leqslant 5$.

For large values of v_i, $F(|v|) \approx \dfrac{-j\ \text{EXP}\ -j\frac{\pi}{2} v_i^2}{\pi |v_i|}$.

Performing the summation, one finds

$$F_{Total}(v) = \frac{-j}{\pi} \sum_{i=1}^{5} \frac{\text{EXP}\ -j\frac{\pi}{2} v_i^2}{|v_i|}. \tag{3.21}$$

Real power attenuation is given by

$$\frac{E_{Total}}{E}^{2} = \frac{F_{Total}^2(v)}{2}. \tag{3.22}$$

Substituting for $F_{Total}(v)$ gives

$$\frac{E_{Total}}{E_o}^{2} = \frac{1}{2\pi^2} \quad \frac{\text{EXP}\ -j\frac{\pi}{2} v_i^2}{|v_i|}^{2}$$

Attenuation due to the fence is $10 \log (E_{Total}/E_o)^2$. The above equation was solved for a distant source of interference, an antenna-to-fence distance of 370 wavelengths and the five fence edge heights ranging from 20.6 to 31.8 wavelengths. Clutter angle was varied from −1 to +1 degrees in 1/4 degree increments and the result is plotted in Figure 3.14. For comparison, the attenuation from a straight-edge fence of the same dimensions is also plotted in Figure 3.14. Considerable improvement in performance can be

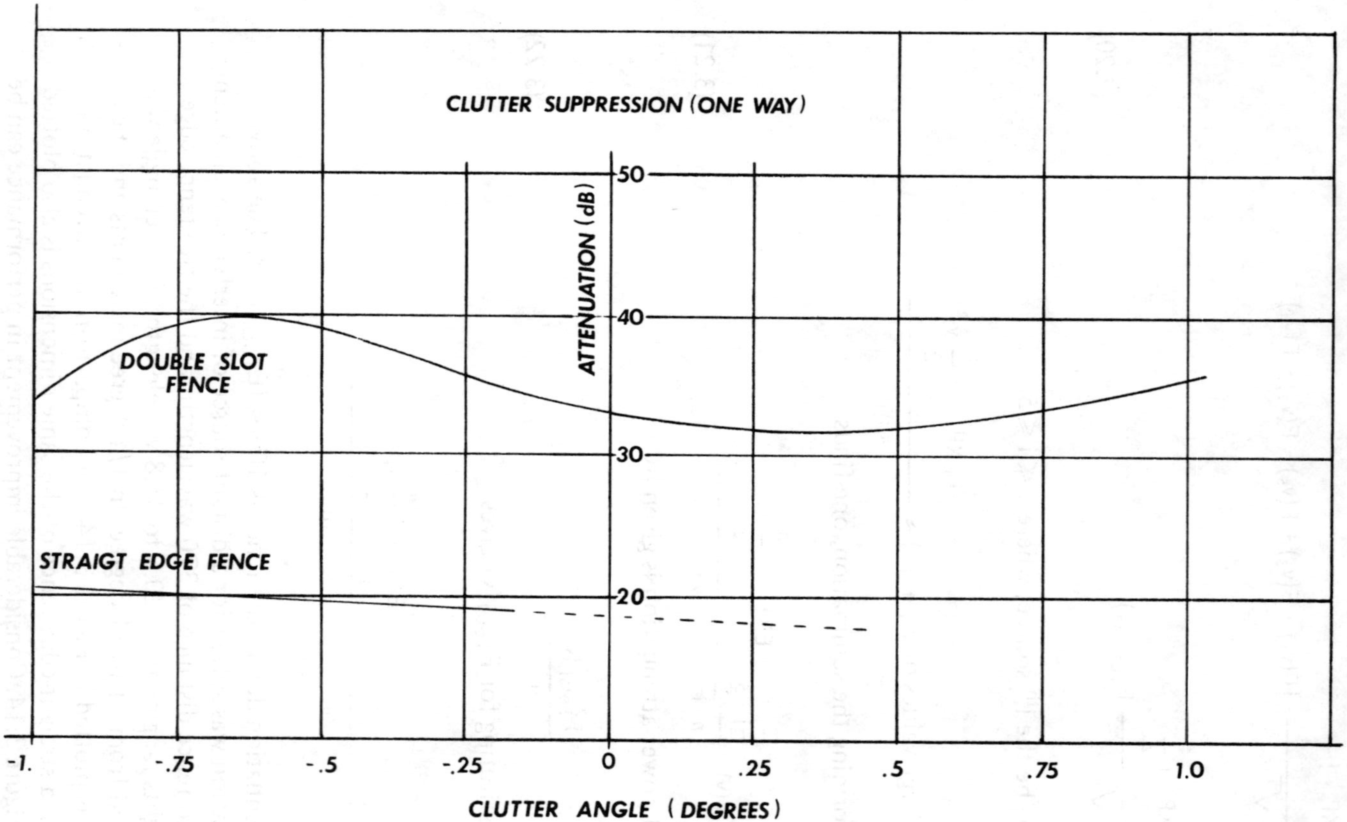

Figure 3.14 Clutter suppression of a double-slot fence as compared to a straight-edge fence

obtained from the double-slot design compared to the straight-edge fence. However, this improvement is true only over a narrow range of clutter angles. Thus, if the source of interference is fixed relative to the receiver a double-slot fence may be used to good advantage.

Becker and Millet [18] have built and tested a double-slot fence and some of their results follow. One reason for building a model was to determine the effect of polarization, which is not taken into account by Fresnel diffraction theory.

Figure 3.15 illustrates their setup for measuring the diffracted field in the shadow region of the fence.

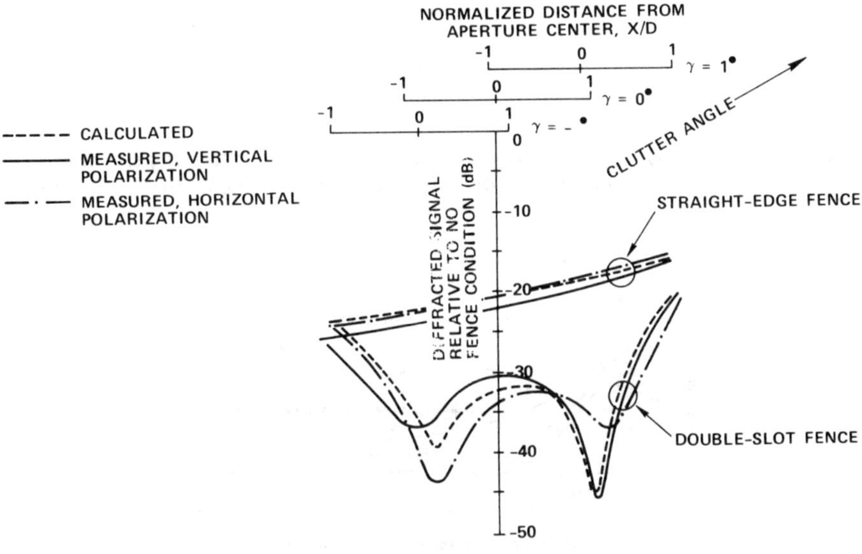

Figure 3.15 Diffracted signal (one-way) from double-slot and straight-edge fences.

According to Becker & Millett:

> The diffracted field in a vertical plane behind the double-slot fence has been calculated and subsequently measured for both vertical and horizontal polarization; the results are shown in Figure 3.15. The diffracted signal is plotted in dB relative to the signal with no fence present; the horizontal scale is distance from the aperture center, normalized to the aperture radius. The horizontal scale slides with angle of the incident clutter; it is shown for three values of clutter angle. The diffracted field behind a straight-edge fence with its edge at the same height at the top edge of the double-slot fence is shown for comparison. The agreement between calculated and measured results is excellent, indicating that Fresnel theory is adequate for this application. The effect of polarization is seen to be a

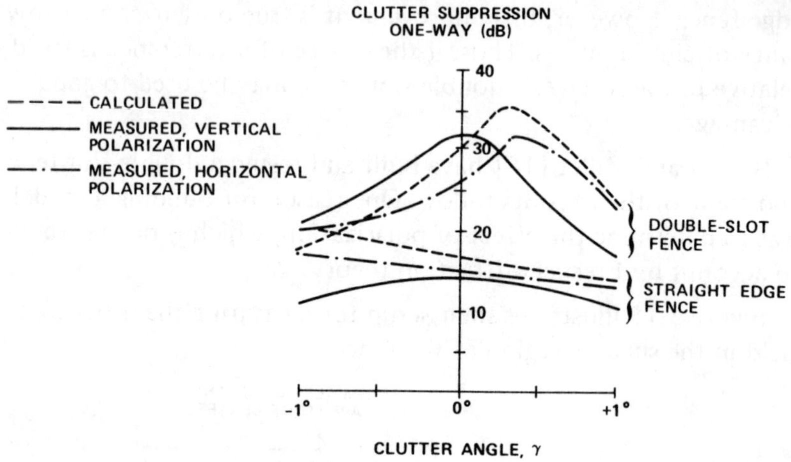

Figure 3.16 One-way clutter suppression as a function of clutter angle. Note: Suppression is average over 10°-20° elevation angles for main beam

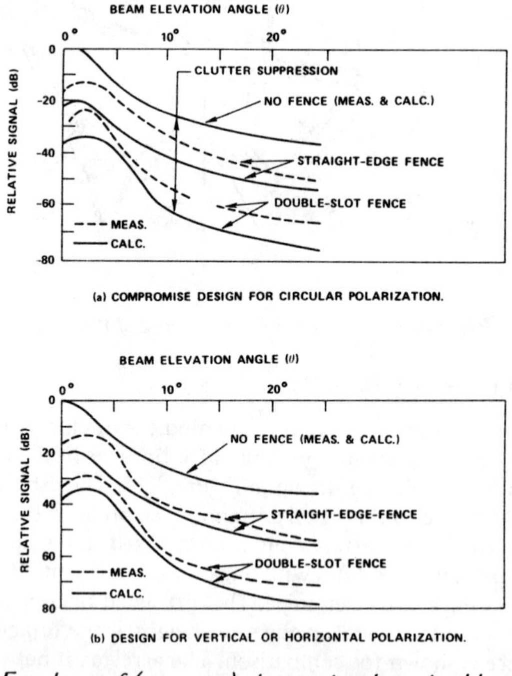

Figure 3.17 Envelope of (one-way) clutter signal received by antenna $\nu = 0°$

slight detuning to either side of the design clutter angle. It is also seen that the diffracted field behind the double-slot fence is suppressed relative to that behind the straight-edge fence by an average of more than 17 dB across the antenna aperture at the design clutter angle (γ - 0°) (see Figure 3.16); furthermore, the field has the desired tapered distribution. For clutter angles other than the design angle, the field distribution deteriorates from the desired one, and the clutter signal suppression decreases; as a result, the total signal received by the large aperture for both positive and negative clutter angles is increased relative to the signal for 0° clutter angle.

Figure 3.17 shows the envelope of the relative signal received by the large aperture of the design frequency and clutter angle as a function of angle beam-pointing angle. The results of calculations and measurements are shown for three cases: no fence, straight-edge fence, and double-slot fence. Clutter suppression is defined as the difference at a given beam-pointing angle between the signal received with the fence and the "no-fence" case.* To simplify presentation of the measurements, the results have been normalized so that the no-fence measured and calculated responses agree completely.

Figure 3.17(a) presents the performance of the fence (dimensions in Table VI) designed for the case of a circularly polarized antenna and circularly polarized clutter signal; the measured results represent a compromise in performance for the two linear components of the circular polarization. As can be seen, the improvement in performance of the double-slot fence over the straight-edge fence is substantiated; the agreement between calculations and measurements is not exact because of the effect of polarization and imperfect simulation of the antenna. The clutter suppression is seen to vary with beam-pointing angle; at 15° above horizontal, the measured one-way clutter suppression from the double-slot fence yields 19 dB improvement over the straight-edge (the calculated improvement is 22 dB). At angles less than 15°, the clutter suppression is somewhat less; it is slightly greater for larger angles.

---

*It is significant to note that the resulting clutter suppression has a significant variation with antenna beam-pointing angle. When the beam is pointed at the top of the fence, all of the diffracted field incident on the aperture is received in phase; for the same beam angle with no fence present, the fields directly incident fron the clutter source are not

Table 3.6
Height of Double-Slot Fence Edges Above Aperture Center

Edge	*Height* (H/λ)
1	20.6
2	22.0
3	27.0
4	30.1
5	31.8

In addition to the calculations and measurements discussed above, the effect of frequency on the double-slot performance was investigated. It was found that while the clutter angle for peak-clutter suppression for a given design varied with frequency, the average clutter suppression over a 2° range of clutter angles remained nearly constant over a ten percent band of frequencies. Here again measurements verified the calculations.

Figure 3.17(b) shows the performance of a fence whose dimensions have been modified slightly to optimize for either horizontal or vertical polarization. It can be seen that for these cases the double-slot fence yields a measured performance which is closer to that calculated. The suppression obtained is several dB improvement over the clutter suppression of a fence designed as a compromise between the two polarizations.

The results shown in Figure 3.14 are somewhat different from those given by Becker and Millett because a simpler model was used which, among other things, did not take into account antenna aperture.

In summary, then, it has been shown that even a simple straight-

---

received in phase. Thus, while the diffracted clutter fields are significantly weaker than the direct clutter, the full improvement is not realized for the condition where the antenna beam is directed at, or near, the top of the fence. The assumption made by Ruze et al. [19] that the clutter suppression for a straight-edge fence is approximately equal to the attenuation of the fields at the center of the antenna aperture can be used only where the antenna beam is pointed considerably above the top of the fence. For quick estimates when the beam is pointed near the fence edge, the result derived only from consideration of the fields can be modified by taking into account the ratio of the antenna pattern gain in the direction of the fence compared to the gain in the direction of the clutter source with no fence present.

edge fence can give considerable decoupling from nearby sources of interference if properly designed. The amount of attenuation is limited only by the maximum physical dimensions allowed for the fence. A double-slot fence is a much better choice than the straight-edge fence if the interference is limited spatially and if the fence can be designed for maximum discrimination in the direction of interference.

For shipboard use, the fence might take the form of a radial platform blocking the line-of-sight of two mutually interfering radars. In this case, it might find functional use as a platform as well as a fence. A more compact form of the shipboard fence might be a small radial platform with a short vertical fence around the circumference. If the source of interference is limited to one direction perhaps a short "picket fence" would suffice.

A fence may be either a solid metal plate or a double wire mesh type, if that is more convenient, unless wind loading is a problem. Additional fence edge treatments are possible and further investigations in this direction should be undertaken.

REFERENCES

[1] Morris, F.J., et al., *Measurement Techniques for Obtaining Transmitter Spectral Output,* RADC-TDR-62-311, 1 December 1961.

[2] Prout, Polisky & Oliva, "Empirical Formula for the Prediction of Microwave Antenna Coupling", IEEE Symposium on EMC, July 1968.

[3] Becker, J.E., *Design Study for AMRAD Anti-Clutter Fence,* Lincoln Laboratories, M.I.T., Report 1233, July 1964.

[4] *A Sphere Technique for the Measurement of Three-Dimensional Radiation Patterns,* The Electro-Mechanics Co., 10 May 1960.

[5] Cain, F.L., K.G. Byers, Jr., *Statistical Gain Characteristics of Radar Antennas,* Final Engineering Report, Project A-958.

[6] Becker, J.E., and J.C. Sureau, "Control of Radar Site Environment by Use of Fences," *I.E.E.E. Antennas and Propagation,* Nov. 1966.

[7] Becker, J.E., and R.E. Millett, "A Double-Slot Radar Fence for Increased Clutter Suppression," *IEEE Trans. Antennas and Propagation,* January 1968.

[8] Mumford, W.W., "Some Technical Aspects of Microwave Radiation Hazards," *Proceedings of the IRE,* Vol. 49, 1961.
[9] Becker, J.E., "Design Study for AMRAD Anti-Clutter Fence," Report 1233, Wheeler Lab. AD 455973.
[10] Hannan, P.W., "The Cassegrain Antenna for the Nike-Hercules Track Radar (U), "*6th Annual Radar Symp. Rec.* (Confidential) (University of Michigan, Ann Arbor, Michigan, June 1960) AD 318-072.
[11] Takada, M., and M. Hinji, "An Application of the Diffraction Grating to the 11-Gc/s Microwave Systems," *IEEE Trans.,* AP-13, July 1965.
[12] Albersheim, W.J., "Radar Interference Reduction; Clutter Fences; Error Analysis; Target Tracking," *IEEE Trans. Aerospace,* Nov. 1967.
[13] Born and Wolf, *Principles of Optics,* New York: Pergamon Press, 1959.
[14] Ruze, J., F.I. Sheftman, and D.A. Cahlander, "Radar Ground-Clutter Shields," *Proceedings IEEE,* Vol. 54, Sept. 1966.
[15] Sureau, J.C., "An Evaluation of the Results of the Theories of Diffraction from a Perfectly Conducting Half-Plane," Wheeler Laboratories Report 1222 P, September 1965.
[16] Preikshat, "Screening Fences for Ground Reflection Reduction," *Microwave Journal,* August 1964.
[17] Becker, J.E., *Design Study for Amrad Anti-Clutter Fence.*
[18] Becker, J., and R. Millett, "A Double-Slot Radar Fence for Increased Clutter Suppression," *IEEE Trans. Antennas and Propagation,* January 1968.
[19] Reference [2] in Theory of a Straight Edge Fence.
[20] McCue, J.J.G., "Remarks on Anti-RFI Fences," *IEEE Transactions on Electromagnetic Compatibility,* Vol. EMC-23, No. 1, February, 1981.

Figure 3.1 Mumford, W.W., "Some Technical Aspects of Microwave Radiation Hazards," *Proceedings of the IRE,* Vol. 49, 1961.

Figures 3.14 to 3.17 Becker, J., and R. Millett, "A Double-Slot Radar Fence for Increased Clutter Suppression," *IEEE Trans. Antennas Propag.,* January 1968.

4

CORRUGATED HORN

Corrugated Horn

A corrugated horn is a modified pyramidal horn that offers significantly improved performance over a standard horn. The modification to the horn takes the form of corrugations in the planes that are perpendicular to the E-field vector. The edges in these planes are responsible for the backlobe radiation of a pyramidal horn. The use of corrugations results in a capacitive surface reactance that does not support the propagation of a surface wave. Without a surface wave much of the illumination of the E-plane edges is eliminated, along with the backlobe radiation.

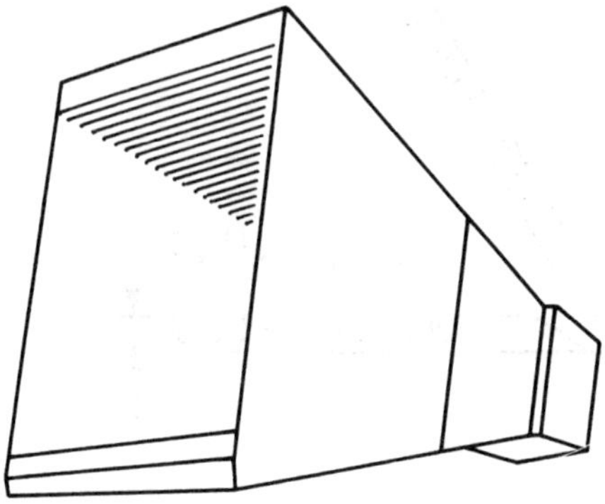

Figure 4.1 Corrugated Horn

To insure fairly broadband operation (two-to-one frequency band) the corrugations or slots should be 10 or more per wavelength. The analysis [1] is simplified when the following assumptions are made: (1) The slot walls are vanishingly thin and (2) only the TEM mode in the slots is reflected from the base of the slots. The higher order modes are attenuated.

The second assumption is the same as requiring that the slot width, g (see Figure 4.2), be small compared with the free-space wavelength and the slot depth d. For such a surface, Elliot [2] showed that the surface reactance can be approximated by

$$X = \frac{g}{g+t} \quad \frac{\mu}{\epsilon} \tan K_o d$$

K_o = propagation coefficient for free space

t = slot wall thickness

when

$$\frac{g}{g+t} \approx 1. \tag{4.1}$$

This condition is met if $t \leqslant g/10$ and the second assumption is satisfied when $g \leqslant \lambda/10$.

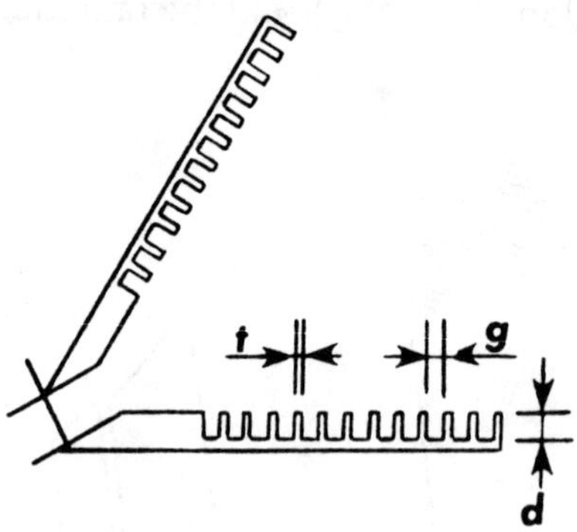

Figure 4.2/Geometry of corrugated horn

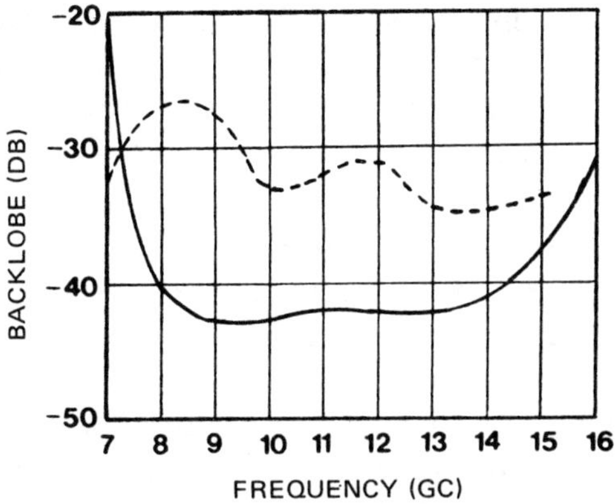

Figure 4.3 Ratio of backlobe to mainlobe levels of the small corrugated horn and the control horn

Since the surface reactance must be capacitive so that a surface wave won't be supported, this requires that $\lambda/4 < d < \lambda/2$ from (1). Hurd [3] has shown that the cutoff depth d_c is partially dependent on the slot width g. However, curves given by Hurd show that the cutoff region is approximately $\lambda/4 < d < \lambda/2$ when $g < \lambda/10$. A small horn was constructed [1] that had a 3½″ square aperture and a 50° flare angle. It had 15 slots per wavelength at

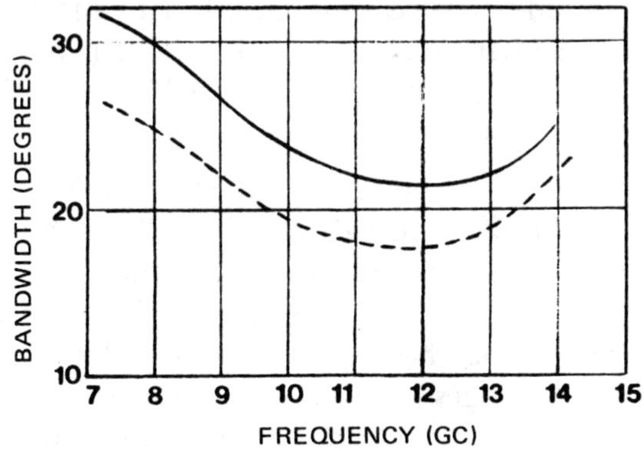

Figure 4.4/3 dB beamwidths.

10 GHz and the slots were λ/4 deep at 8 GHz or the lower end of its frequency range. The control horn is the same horn but with the slots covered with aluminum tape and paint. The curves of backlobe level versus frequency for both horns are given in Figure 4.3.

The 3 dB beamwidth versus frequency curves of the two small horns are given in Figure 4.4. As can be seen from Figure 4.4, there is a slight widening of the 3 dB beamwidth but, as can be seen from Figure 4.5, had a 10 dB beamwidth requirement been chosen, the corrugated horn would have the lower beamwidth. A substantial improvement in the backlobe radiation from the small corrugated horn is evident from Figure 4.5 and this pattern is essentially unchanged for the frequency band of 8 to 14 GHz.

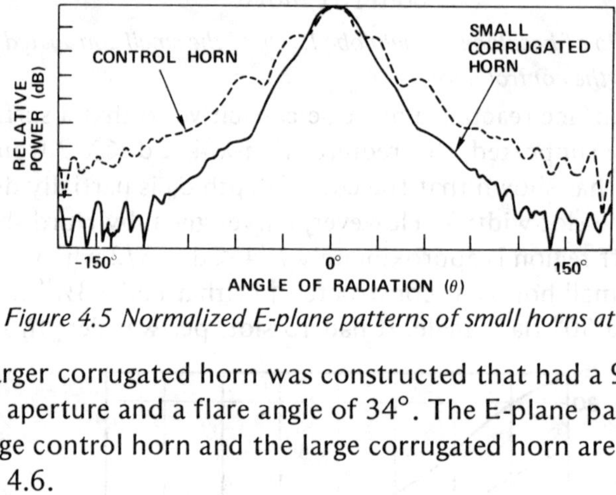

Figure 4.5 Normalized E-plane patterns of small horns at 10 GHz

A larger corrugated horn was constructed that had a 9.7 inch square aperture and a flare angle of 34°. The E-plane patterns for the large control horn and the large corrugated horn are shown in Figure 4.6.

The calculated gain, on bore-sight, of the large corrugated horn is improved by 1 dB to 22.5 dB from the 21.5 dB of the control horn. This change in directivity is the result of the narrowed beamwidth caused by the corrugations. The saddle, which was at the center of the beam and had been caused by the fields being diffracted by the E-plane edges has also been eliminated. The backlobe is seen to be 57 dB down from the main beam; this represents a 27 dB improvement.

The corrugations should be kept a short distance away from the waveguide feed so as to not affect the impedance of the antenna. With the corrugations starting 2.5″ from the waveguide feed, the

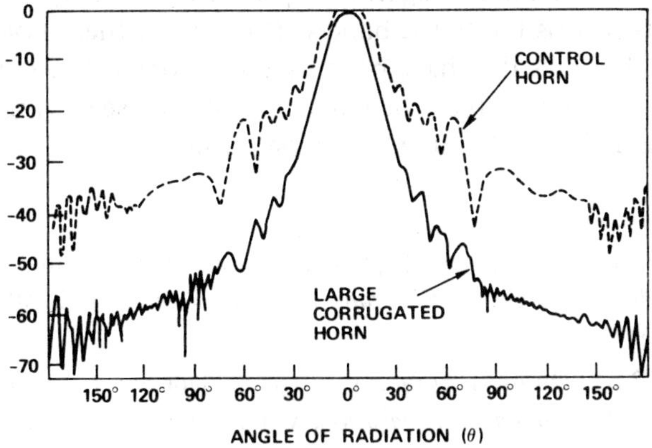

Figure 4.6 E-Plane patterns of large control and corrugated horns at 10 GHz

VSWR of the large corrugated horn was measured over a range of frequencies. The result is plotted in Figure 4.7.

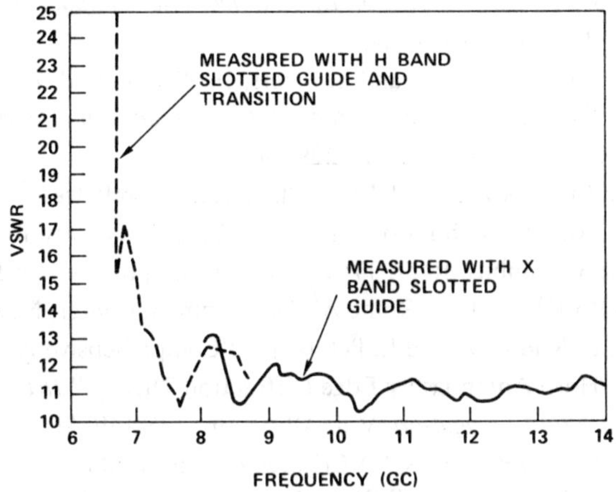

Figure 4.7 VSWR of the large corrugated horn

Conclusions

The use of corrugations or slots in pyramidal horns has been shown to be an effective way of reducing undesirable backlobe radiation. This reduction is limited by the amount of direct, line-of-sight illumination of the top and bottom edges of the horn from the waveguide.

A relatively constant pattern over a wide frequency range can be obtained if the horn flare angle is sufficiently large. The corrugated wall impedance must be capacitive for hybrid mode operation, requiring that the corrugation depth be between 1/4 λ and 1/2 λ. In effect, this limits the bandwidth ratio of these horns to 1.7 to 1. Techniques that have been used or proposed to extend the capacitive bandwidth of the corrugations are use of ridge loaded slots, dielectric loading, and tapered slots.

REFERENCES

[1] Laurie, R.E., and L. Peters, Jr., "Modifications of Horn Antennas for Low Sidelobe Levels," *IEEE Trans. Antennas Propag.,* Vol. AP-14, September 1966.

[2] Elliott, R.S., "On the Theory of Corrugated Plane Surfaces," *IRE Trans. on Antennas and Propagation,* Vol. AP-2, pp. 71-81, April 1954.

[3] Hurd, R.A., "The Propagation of an Electromagnetic Wave Along an Infinite Corrugated Surface," *Canadian Journal of Physics,* Vol. 32, pp. 727-734, Dec. 1954.

[4] Peters, L., Jr., and R.C. Rudduck, "RFI Reduction by Control of Antenna Sidelobes," *IEEE Trans. on Electromagnetic Compatibility,* Vol. EMC-6, pp. 1-11, January 1964.

[5] Russo, P.M., R.C. Rudduck, and L. Peters, Jr., "A Method for Computing E-Plane Patterns of Horn Antennas," *IEEE Trans. on Antennas and Propagation,* Vol. AP-13, pp. 219-229, March 1965.

[6] Yu, J.S., and R.C. Rudduck, "The H-Plane Radiation Pattern of Horn Antennas," Antenna Laboratory, The Ohio State University Research Foundation, Columbus, Rept. 1767-5, May 1965, prepared under Contract AF 30(602)-2711 for Rome Air Development Center, New York.

[7] Yu, J.S., R.C. Rudduck, and L. Peters, Jr., "Comprehensive Analysis for E-Plane Horn Antennas by Edge Diffraction Theory," *IEEE Trans. on Antennas and Propagation,* Vol. AP-41, pp. 138-149, March 1966.

[8] Sciambi, A.F., and P. Foldes, "A Critique of Advanced High Performance Antenna Feed Systems," RCA Missile and Surface Radar Division, Moorestown, N.J., 1964.

[9] Peters, L., Jr., and R.C. Rudduck, "Application of Electromagnetic Absorbing Materials as Interference Reduction Techniques," Antenna Laboratory, The Ohio State University, Research Foundation, Columbus, Rept. 1423-7, November 1963, prepared under Contract AF 30(602)-2711 for Rome Air Development Center, Air Research and Development Command, Griffiss AFB, New York.

[10] Kritikos, H., R. Dresp, and K.C. Lang, "Studies of Antenna Sidelobe Reduction," Rept. RADC-TDR-64-355, Vol. 1, October 1964, Rome Air Development Center, Griffiss AFB, New York.

[11] Kay, A.F., "The Scalar Feed", TRG Inc., East Boston, Mass., March 30, 1964. Sci. Rept. 5, prepared under Contract AF 19(604)-8057, Air Force Cambridge Research Laboratories, Office of Aerospace Research, USAF, Bedford, Mass.

[12] Rhodes, D.R., "An Experimental Investigation of the Radiation Patterns of Electromagnetic Horn Antennas," *Proc. IRE,* Vol. 36, pp. 1101-1105, September 1948.

[13] Schelkunoff, A.S., *Electromagnetic Waves.* Princeton: Van Nostrand, 1964.

[14] Peters, L., Jr., "Passive Bistatic Enhancement Devices," *Proc. IEEE* (London), July 1961.

[15] Potter, P.D., "A New Horn Antenna With Suppressed Side Lobes and Equal Beamwidths," *Microwave Journal,* June 1963.

[16] Jakes, W.C., Jr., "Horn Antennas," in *Antenna Engineering Handbook,* H. Jasik, Ed. New York: McGraw-Hill, 1961.

[17] Lyon, J.A.M., C.J. Digenis, W.W. Parker and M.A. Ibrahin, "Electromagnetic Coupling Reduction Techniques," Radiation Laboratory, University of Michigan, June 1968.

[18] K.M. Chen and V.V. Liepa, "The Minimization of Backscattering of a Cylinder by Central Loading," *IEEE Trans. on Antennas and Propagation,* Vol. AP-12, 1964.

[19] Fernando, W.M.G., and H.E.M. Barlow, "An Investigation of the Properties of Radial Cylindrical Surface Waves Launched Over Flat Reactive Surfaces," *Proc. IEEE,* Vol. 103, 1956.

[20] Rotman, W., "A Study of Single Surface Corrugated Guides, "*Proc. IRE,* Vol. 39, 1951.

[21] Senior, T.B.A., "Control of the Acoustic Scattering Characteristics of a Rigid Sphere by Surface Loading," *Journal Acoustic Soc. Amer.,* Vol. 37, 1965.

[22] LaGrone, A.H., and G.F. Roberts, "Minor Lobe Suppression in a Rectangular Horn Antenna through the Utilization of a High Impedance Choke Flange," *IEEE Trans. on Antennas and Propagation,* Vol. AP-14, January 1966.

[23] Bahret, W.F., and L. Peters, Jr., "Small-Aperture Small-Flare Angle Corrugated Horns," *IEEE Trans. on Antennas and Propagation,* Vol. AP-16, July, 1968.

[24] Peters, L. Jr., and T.E. Kilcoyne, "Radiating Mechanisms in a Reflector Antenna System," *IEEE Trans. on Electromagnetic Compatibility,* Vol. EMC-7, Dec. 1965.

[25] Kuo, S.C., and C.G. Roberts, "Broad Band Corrugated Horns," IEEE International Symposium on Antennas and Propagation, 1975.

Figures 4.3 to 4.7 Laurie, R.E., and L. Peters, Jr., "Modifications of Horn Antennas for Low Sidelobe Levels," *IEEE Trans. Antennas Propag.*, Vol. AP-14, September 1966. © 1966 IEEE.

5

DIELGUIDE

GENERAL DESCRIPTION OF THE DIELGUIDE

The Dielguide is a feed modification typically used in Cassegrain antennas which improves antenna performance, i.e., efficiency. The Dielguide development is an outcome of the antenna designer's need for a more efficient, low noise antenna. This development is a step closer to the ideal of a uniformly illuminated reflector with no spillover, which maximizes the antenna's gain-to-noise ratio. Other developments [1, 2], although partially successful, have resulted in complicated feeds that are narrowband and impractical to implement.

Conventional aperture antenna design is a compromise between uniform illumination of the reflector and spillover past the subreflector. This is because the radiation pattern of the conventional feed varies with angle in a manner which is described by diffraction theory. As a result of this compromise, the conventionally designed Cassegrain antenna has an efficiency of about 55 per cent. Use of the Dielguide feed can increase efficiency to around 75 per cent.

In a Cassegrain antenna, the Dielguide is a dielectric structure placed between the feed horn and the subreflector. It also supports the subreflector and therefore eliminates the need for struts, which are undesired reflectors and diffractors of main beam radiation.

The purpose of the Dielguide is to guide radiation from the feed to the subreflector.

The Dielguide is both conceptually simple and has the added advantage of being a broadband device.

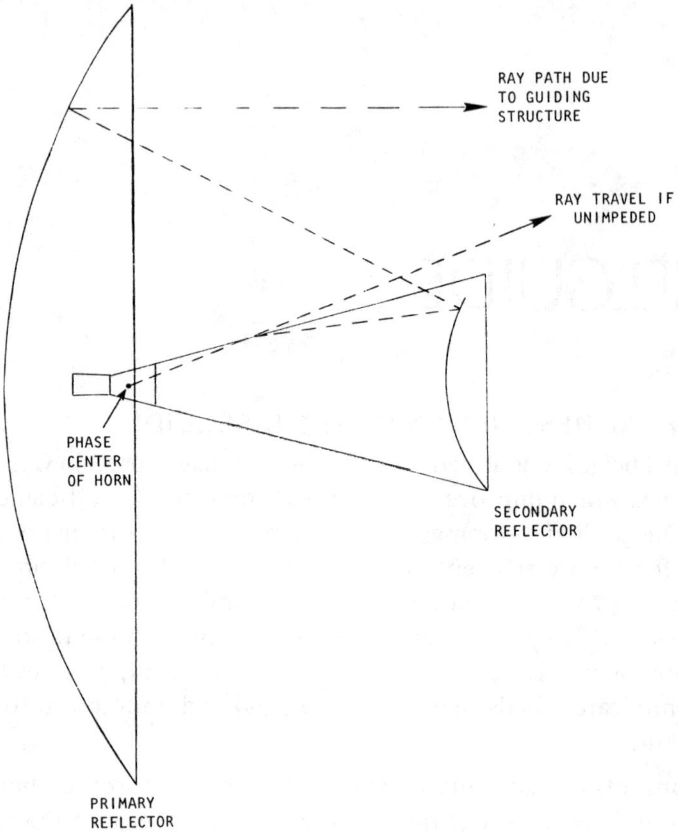

Figure 5.1/Cassegrain antenna utilizing dielguide feed

THEORY OF OPERATION OF THE DIELGUIDE

The Dielguide works by reflecting energy at the dielectric-air boundary back to the subreflector which otherwise would miss it. The basic principle is illustrated in Figure 5.1. Rays that would be radiated past the secondary reflector in "spill-over lobes" as shown in Figure 5.2, encounter the dielectric material.

When a plane wave falls onto a boundary between two homogeneous media of different electrical properties, it is split into two

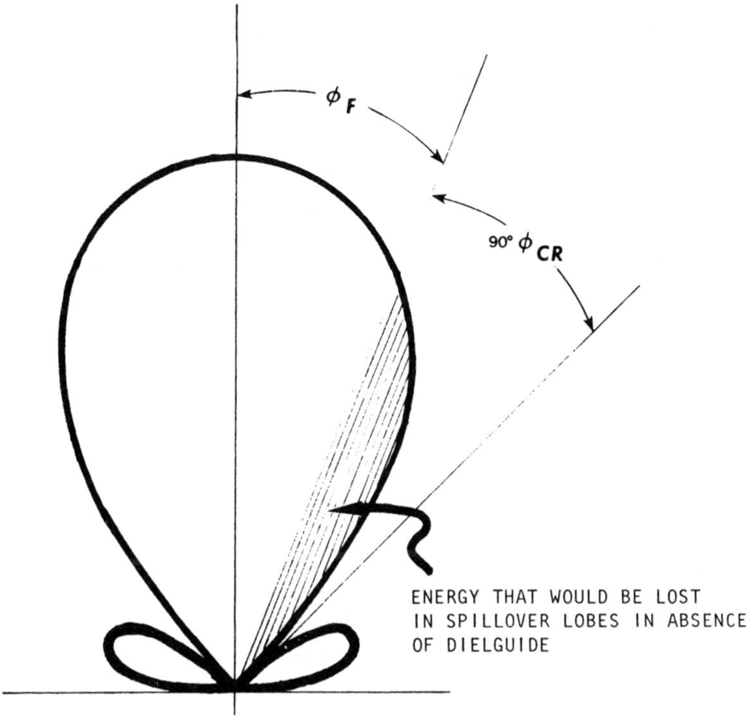

Figure 5.2/Effect of guiding structure on radiation pattern of metal horn.

waves; a wave proceeding into the second medium and a reflected wave propagated back into the first medium. A plane wave propagated in the direction specified by the unit vector $\vec{S}_i$ (i indicating incident wave) is completely determined when the time behavior at one particular point in space is known. For if F(t) represents the time behavior at any one point, the time behavior at another point, whose position vector relative to the first point is $\vec{r}$, is given by $F(t - (\vec{r} \cdot \vec{S})/v)$ where v is the velocity of propagation. At the boundary between the two media, t, the time variation of the secondary fields, will be the same as that of the incident primary field. Hence, if $\vec{S}_r$ and $\vec{S}_t$ denote unit vectors in the direction of propagation of the reflected and transmitted wave, one has, on equating the arguments of these wave functions at a point r(x,y,z) on the boundary:

$$t - \frac{\vec{r} \cdot \vec{S}_i}{v_1} = t - \frac{\vec{r} \cdot \vec{S}_r}{v_1} = t - \frac{\vec{r} \cdot \vec{S}_t}{v_2} \quad (5.1)$$

where v_1 and v_2 are the velocities of propagation in the two media. If we take the boundary as the plane z = 0,

$$\frac{x\,s_{ix} + y\,S_{iy}}{v_1} = \frac{x\,S_{rx} - y\,S_{ry}}{v_1} = \frac{x\,S_{xt} + y\,S_{yt}}{v_2} \quad (5.2)$$

Since this must hold for all values x and y on the boundary,

$$\frac{S_{ix}}{v_1} = \frac{S_{rx}}{v_1} = \frac{S_{tx}}{v_2}, \quad \frac{S_{iy}}{v_1} = \frac{S_{ry}}{v_1} = \frac{S_{ty}}{v_2} \quad (5.3)$$

The plane specified by S_i and the normal to the boundary, Figure 5.3, is the plane of incidence. Therefore,

$$S_{ix} = \sin\theta_i, \qquad S_{iy} = 0, \qquad S_{iz} = \cos\theta_i$$
$$S_{rx} = \sin\theta_r, \qquad S_{ry} = 0, \qquad S_{rz} = \cos\theta_r$$
$$S_{tx} = \sin\theta_t, \qquad S_{ty} = 0, \qquad S_{tz} = \cos\theta_t$$

and

$$\frac{\sin\theta_i}{v_1} = \frac{\sin\theta_r}{v_1} = \frac{\sin\theta_t}{v_2}.$$

Hence,

$$\sin\theta_i = \sin\theta_r$$

$$\cos\theta_r = -\cos\theta_i$$

$\theta_r = -\theta_i$, which is the law of reflection.

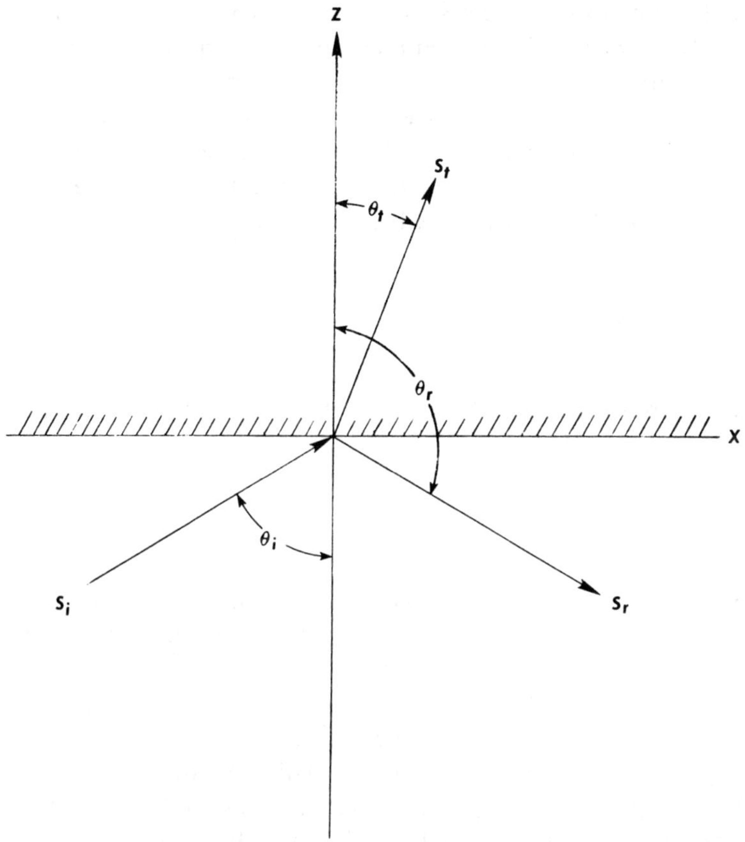

Figure 5.3 The plane of incidence

Using Maxwell's relation between the refractive index and the dielectric constant,

$$\frac{\sin\theta_i}{\sin\theta_t} = \frac{v_1}{v_2} = \frac{\epsilon_2\mu_2}{\epsilon_1\mu_1} = \frac{n_2}{n_1} = n_{12} \tag{5.4}$$

This relation is the law of refraction (Snell's law).

When $n_2 > n_1$, then $n_{12} > 1$, and it may be said that the second medium is electrically denser than the first medium. In this case,

$$\sin\theta_t = \frac{1}{n_{12}}\sin\theta_i < \sin\theta_i \tag{5.5}$$

so that there is a real angle θ_t of refraction for every angle of incidence. If, however, the second medium is electrically less dense than the first medium (i.e., if $n_{12} < 1$), one obtains a real value for θ_t only for those incident angles θ_i for which $\sin\theta_i \leqslant n_{12}$. For larger values of the angle of incidence, total reflection takes place, i.e., reflection occurs when energy is propagated from an electrically dense medium into one which is less dense, or

$$n_{12} = \frac{n_2}{n_1} = \sqrt{\frac{\epsilon_2\mu_2}{\epsilon_1\mu_1}} < 1 \tag{5.6}$$

provided that the angle of incidence θ_i exceeds the critical value θ_c given by

$$\sin\theta_c = n_{12}. \tag{5.7}$$

At the critical angle, there is no energy transmitted into the second medium, since the angle θ_t is 90°. As the angle of incidence increases beyond this point, all the energy is reflected back, a condition of total reflection. There is, however, an electromagnetic field established in the second medium even though there is no energy flow across the boundary [3]. This field decreases rapidly with the depth of penetration, the effective depth being on the order of a wavelength; at 7 GHz, this would be 4 centimeters. There is a way of verifying the existence of this field [4]. The interpretation of this phenomenon is that while there is no lasting flow of energy into the reflecting medium, energy flows to and fro.

RESULTS AND TRADE-OFF CURVES

Most Dielguide materials used to date have been polystyrene based plastic foams. In these cases, the dielectric constant is between 1.08 and 1.5 and the loss tangent has been 0.0004 or less. In the experimental models fabricated, the loss due to the dielectric material was less than 0.1 dB. This is the same as a noise temperature of 7°K maximum which is more than made up for by the noise temperature reduction due to less spillover energy.

Figure 5.4 shows a Cassegrain Dielguide for monopulse tracking

Figure 5.4/Four reflector array of Cassegrain Dielguides

purposes. The subreflector is attached directly to the large end of the Dielguide.

The Dielguide allows use of a smaller subreflector as follows. In a conventional Cassegrain feed, the subreflector illumination will be tapered to about −10 dB to −12 dB at the edges and the subreflector size is used to control the slope of the main pattern skirts. The use of the Dielguide, however, produces a subreflector pattern that has very steep skirts. The subreflector can now intercept the first few nulls of the feed pattern and still remain small, on the order of ten wavelengths. Now, not only does the subreflector reflect more energy to the dish but it also transforms its steep pattern skirts into proportionally steep pattern skirts of the main reflector. The effect is about a 2 dB overall gain of the Dielguide Cassegrain as compared with a conventional Cassegrain.

Examples of E, H, and 45° plane patterns of a Cassegrain Dielguide feed are shown in Figure 5.5. These primary patterns were measured across the base of the Dielguide and are therefore the illumination of the subreflector.

These patterns have very steep skirts as a result of energy reflection provided by the dielectric material and they have fairly constant illumination across the beam. Also, the patterns illustrate the circular symmetry needed for high antenna efficiency.

The corresponding main beam or scatter patterns are shown in

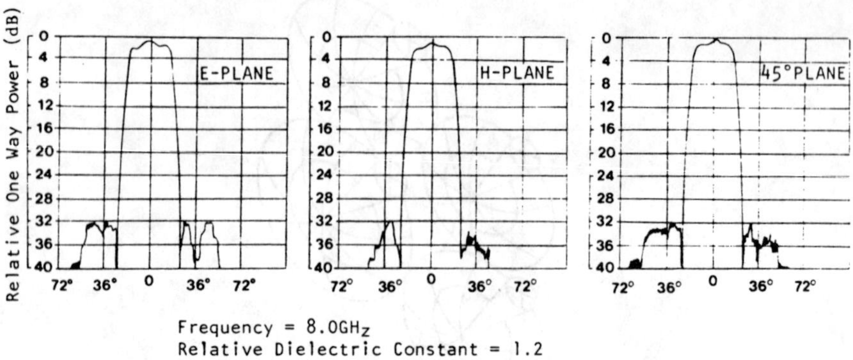

Figure 5.5 Typical primary pattern – Cassegrain Dielguide feed

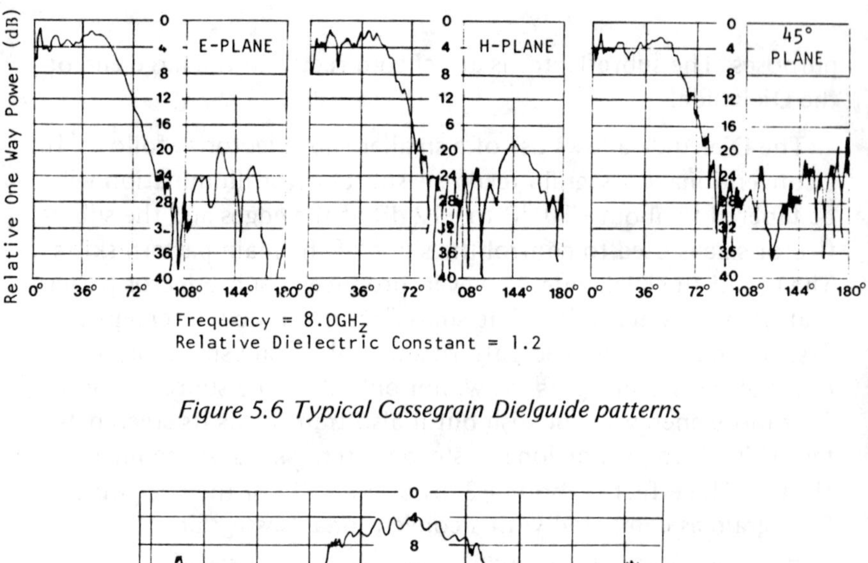

Figure 5.6 Typical Cassegrain Dielguide patterns

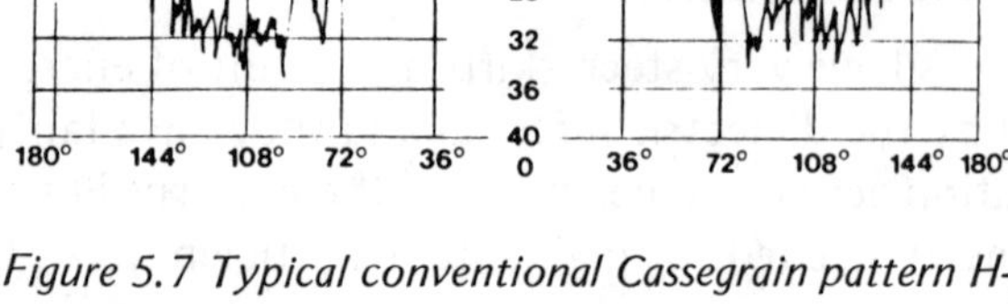

Figure 5.7 Typical conventional Cassegrain pattern H-plane

Figure 5.6. The resulting patterns of Figure 5.6 have uniform illumination, fairly steep skirts and reduced first sidelobes. The circular symmetry is also apparent.

For comparison, the radiation pattern from a conventionally fed, 40 wavelength subreflector Cassegrain antenna is shown in Figure 5.7. The 3 dB beamwidth illumination is not as uniform as that of Figure 5.6 and the backlobes are much higher.

The antenna whose pattern is given in Figure 5.7 had a subreflector diameter of 5 feet and operated at a frequency of 8.85 GHz. The scatter pattern of Figure 5.6 has a more uniform illumination and lower first sidelobes than does the pattern of Figure 5.7. The conventional Cassegrain had a 40 wavelength subreflector as compared to a 10 wavelength subreflector for the Dielguide Cassegrain.

Table 5.1 gives some predicted aperture efficiencies as a function of frequency. The aperture efficiency is the product of the spillover efficiency times the illumination efficiency. These efficiencies were calculated from radiation pattern data obtained from a Cassegrain Dielguide that was used in an 8.5 foot diameter main reflector.

TABLE 5.1
Predicted Aperture Efficiency

Frequency	η_{sp}	η_{at}	$\eta_{ap}\eta_{at}$
7.2 GHz	91.0%	96.0%	87.3%
7.8 GHz	94.5%	93.5%	882.%
8.4 GHz	95.3%	94.2%	89.7%

η_{sp} is spillover efficiency
η_{at} is illumination efficiency

As can be seen from Table 5.1, little energy is spilled over, either past the subreflector or past the main reflector. This is important because spillover energy contributes to antenna noise temperature. The predicted aperture efficiency, for this case, is at least 87 per cent as compared with 55 to 65 per cent for a conventional feed. The conventional feed would then result in an overall antenna efficiency of 45 to 55 per cent.

Figure 5.8 gives the radiation pattern for the above mentioned antenna with the 8.5 foot dish.

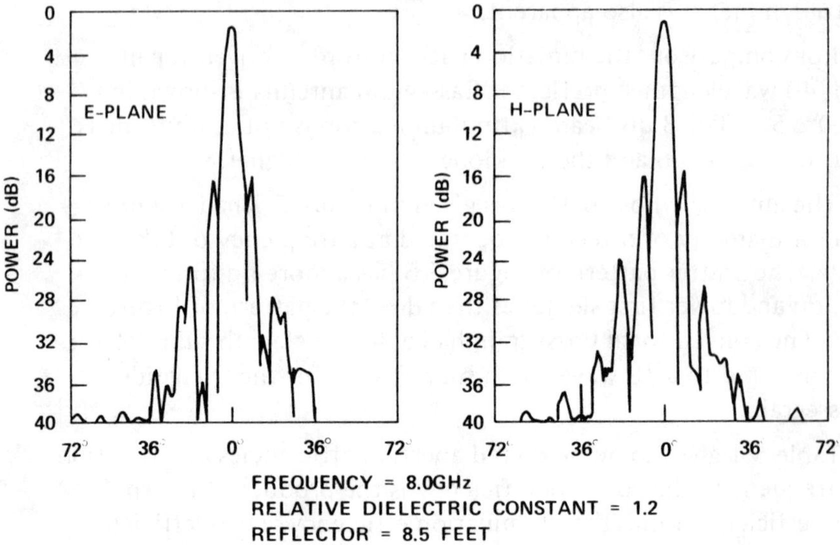

Figure 5.8 Typical radiation pattern Cassegrain Dielguide feed

The measured gain for the antenna pattern in Figure 5.8 was 45.4 dB at 7.8 GHz, which gave an overall antenna efficiency of 75 per cent. The measured efficiency over the 7.2 to 8.4 GHz band was at least 75 per cent.

As previously noted, an additional benefit from using the Dielguide technique is that it is possible to use smaller subreflectors. This allows the formation of Cassegrain apertures with much smaller main reflectors than had previously been possible. The cost of doing so is a reduction in antenna efficiency. For instance, a Dielguide feed with a 3.7 wavelength subreflector resulted in an antenna efficiency of about 65 to 70 per cent. It may be possible to utilize main reflectors as small as 10 wavelengths in diameter. The pattern from an antenna using a 3.7 wavelength subreflector is shown in Figure 5.9.

As far as power capacity is concerned, some Dielguides have been tested to a peak power level of 3 MW at an average value of 4 KW at 5.6 GHz without damage. In another application, quartz wool was used where the power level was 20 KW CW without overheating.

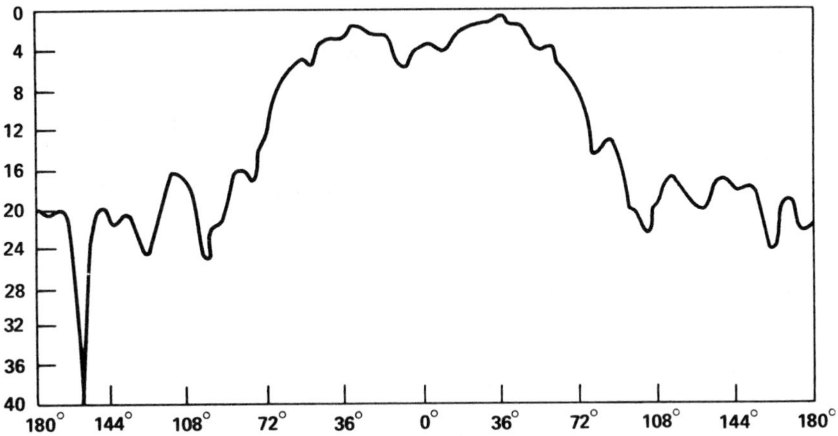

Figure 5.9 Typical Radiation pattern (3.7 λ Subreflector)

Conclusions

The Dielguide is a feed modification to the antenna which results in greater efficiency and a lower noise temperature figure. Efficiencies on the order of 75 to 80 per cent are easily achievable and gain is increased by 1.5 to 2 dB. The technique is not a radical departure from the normal Cassegrain antenna structure and has a fairly high power capacity. The first sidelobes are reduced but the far-out sidelobes are not affected. For this reason, in cases where there exists severe interference coupling between radars, this approach may not help much. It is seen as an improvement in the region of the main beam and first few sidelobes but not elsewhere.

REFERENCES

[1] Potter, F.D., "A New Horn with Suppressed Sidelobes and Equal Beamwidths," *Microwave Journal,* June 1963.

[2] Schuster, D., et al., "The Determination of Noise Temperatures of Large Paraboloidal Antennas," *IRE Trans. PGAP,* May 1962.

[3] Born and Wolf, *Principles of Optics.* Elmsford: Pergamon Press, 1964.

[4] Culshaw and Jones, *Proceedings of the Physical Society,* Vol. B66, 1954.

[5] Ramos, S., and J.R. Whinnery, *Fields and Waves in Modern Radio.* New

York: Wiley and Sons, Inc., 1953.

[6] Ricardi, L.J., and L. Niro, "Design of a Twelve-Horn Monopulse Feed," *IRE Int'l. Conv. Rec.*, Pt. 1, 1961.

[7] Jensen, P.A., "A Low-Noise Multimode Cassegrain Monopulse Feed with Polarization Diversity," *Nerem Rec.*, 1963.

[8] Hannan, P.W., "Optimum Feed for All Three Modes of a Monopulse Antenna II: Practice," *IRE Trans.*, AP-9, 1961.

[9] Final Report – "Dielguide Cassegrain Feed Modification for Satellite Communication Terminal AN/MSC-46," Radiation, Inc., AD 816840.

Figures 5.5 to 5.7 Jensen, P.A., "A Low-Noise Multimode Cassegrain Monopulse Feed with Polarization Diversity," *Nerem Rec.*, 1963.

Figures 5.8 and 5.9 Dielguide Cassegrain Feed Modification for Satellite Communication Terminal AN/MSC-46, Final Report, Radiation, Inc., AD 816840.

6

TUNNEL ANTENNA

The most common form of the tunnel antenna is a reflector type antenna with a tunnel or shroud extending out from the reflector.

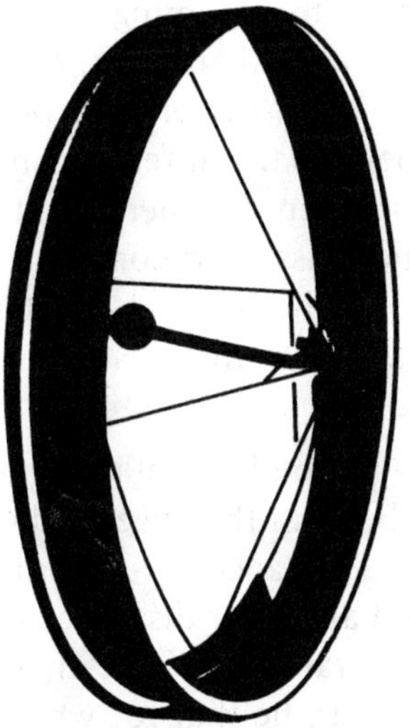

TYPICAL HIGH PERFORMANCE ANTENNA (RADOME REMOVED)

Figure 6.1 Tunnel Antenna

The tunnel is lined with absorbent material to minimize the reradiated, off-axis energy. The object of the tunnel is to reduce the amount of energy spilled over the main dish and subreflector. The side lobe radiation is due entirely to the field illuminating the edges of the tunnel.

The tunnel is placed in such a way around the antenna as to obstruct the antenna's original side lobes (see Figure 6.1). The analysis [1], although not presented here, could utilize diffraction theory [2] to produce calculations of:

(a) The illumination of the front edge of the tunnel;
(b) Diffracted field generated by this discontinuity (this will depend upon the nature of the tunnel wall and edges);
(c) The lobe patterns resulting from the interference between the waves diffracted at various tunnel edges; and
(d) Secondary diffraction occurring at the rear edge of the tunnel.

Edge illumination by the main beam depends on the length and width of the tunnel. The contribution of the feed can only be evaluated accurately by obtaining and analyzing the feed pattern. If the pattern is not available, the estimated magnitude of this contribution can be determined by assuming that the tunnel edge is in the Fraunhofer region of the feed antenna, and that the gain in the direction of the edge is of the order of 8 dB below that of an isotropic source [3]. The field of the antenna can be calculated from an assumed field distribution in the aperture. The phase of the field in the aperture plane is assumed constant; calculations can be made with the amplitude of the field (a) constant over the aperture and (b) varying sinusoidally over the aperture (i.e., zero at the edges, maximum at the center).

Close to the antenna — in the Fresnel region — the antenna produces a flashlight beam (beyond the edge of the beam the field falls off very sharply). At a distance of approximately D^2/λ, the beam begins to spread and at longer distances we deal with the Fraunhofer region, i.e., the far field of the antenna. The choice of tunnel length would appear to be that for which the illumination from the two sources (the main beam of the parabolic reflector and the edge of the tunnel) is about the same. After obtaining an amplitude plot of the field in the neighborhood of the antenna, one can choose by inspection a location for the front edge of the tunnel at which the illumination is low. If the tunnel diameter is chosen then the optimum length can be read off the amplitude plot.

Although there is reduction in side lobe level with increase in length of the tunnel, experimental investigation [4] of a 12 inch disk antenna with a tunnel proves that there is a decrease in the gain of the main beam as the length of the tunnel is increased. Consequently, a relation between the optimum tunnel length and the size of the antenna must be investigated.

One of the major manufacturing sources of tunnel antennas (with and without radomes) [5] is the Andrew Corporation, which developed parabolic reflector antennas in four sizes (8, 10, 12, and 15 feet) with frequencies ranging from 1.7 GHz to 12 GHz, and dual polarization. The lengths of tunnels vary from approximately 2 feet to more than 4 feet for the corresponding reflectors, while the tunnel diameters are slightly larger than the respective apertures.

An example of one of Andrew's high-performance shrouded antennas is given below. The first sidelobes are 30 dB down but the shroud or tunnel has little effect in this region. The near sidelobes (25-35°) are down 50 dB, but the main reduction occurs in the rear 180° sector where the radiation is down at least 65 dB (see Figure 6.2).

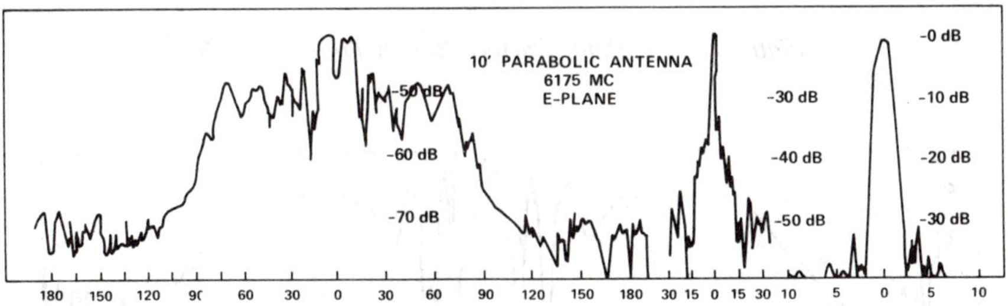

Figure 6.2 Typical E-plane pattern of a tunnel antenna

Another study [6] indicated that sidelobe levels could be reduced to the −40 dB to −50 dB levels in the 60 to 180 degree sector with only a slight loss in gain and slight increase in beamwidth at the lesser tunnel depths. This represents a 10 dB to 25 dB reduction in sidelobe levels. The tunnels were used with parabolic antennas that had reflectors of ten wavelengths or greater in diameter and measurements were taken at 9.375 GHz. At extreme tunnel depths, a reduction of the near sidelobes is possible but only at the expense of gain.

Because of the high angles of incidence found in the operation of a tunnel, the HAIR type absorbers work best. However, they are usually thicker than the resonant or tuned type for the same frequency, and for this reason the resonant types are probably best suited for use in a tunnel. The tunnel should extend around the back of the reflector or a strong back lobe may appear.

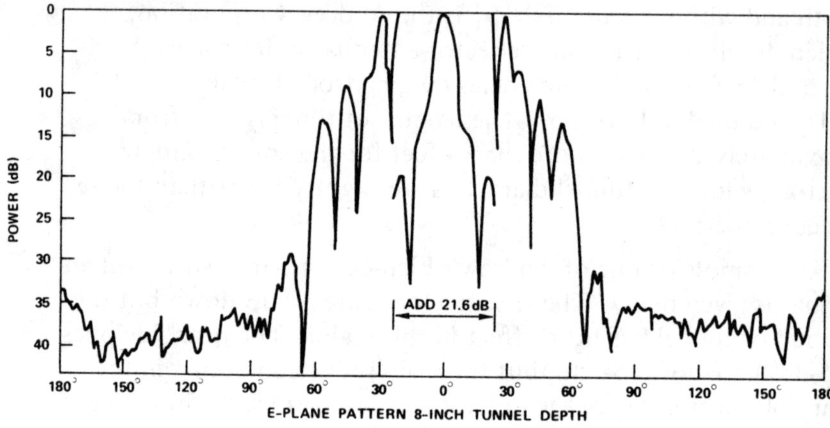

Figure 6.3a E-Plane Pattern 8-inch Tunnel Depth

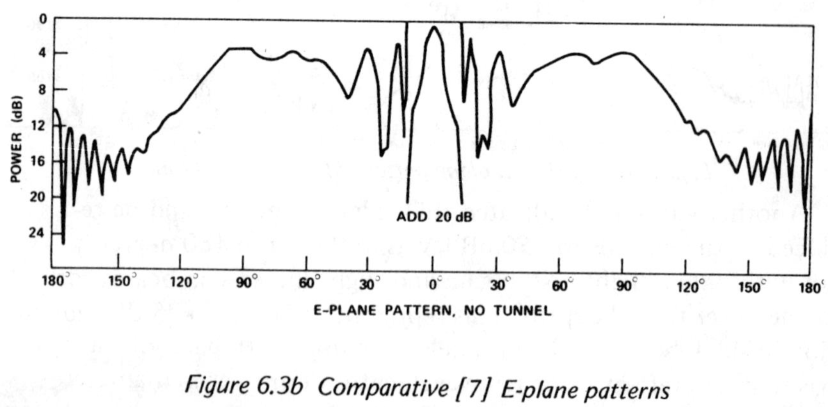

Figure 6.3b Comparative [7] E-plane patterns

Conclusion

By way of summary [6], the following is a representative comparison between an unmodified model of an operational radar and the same model with a tunnel added:

	Unmodified	Tunnel
Main Lobe and Near-In Sidelobes	– Little Change –	
Average −30 dB Level from Maximum	7.5 beamwidths	5 beamwidths
Average −40 dB Level from Maximum	90 beamwidths	21 beamwidths
Average −50 dB Level from Maximum	120 beamwidths	27 beamwidths
Average −60 dB Level from Maximum	–	90° (Andrew HP-10-59)

The use of a tunnel results in a significant reduction, on the order of 10 dB to 25 dB, of the far-out sidelobes. Also, the addition of a tunnel, the hair type absorbers work best. However, they are usually thicker than the resonant or tuned type for the same frequency, and for this reason the resonant types are probably best suited for use in a tunnel. The tunnel should extend around the back of the reflector or a strong back lobe may appear (see Figures 6.3 and 6.4).

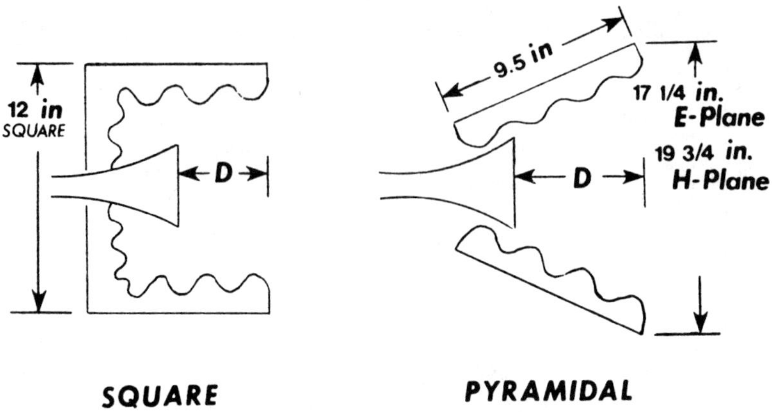

Figure 6.4 Absorber tunnel configurations

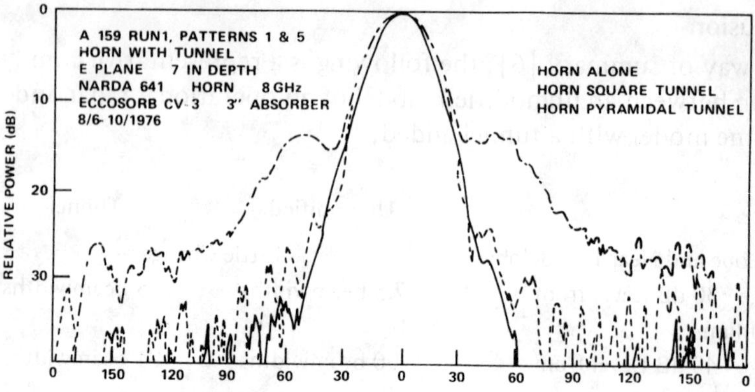

(a) PLANE

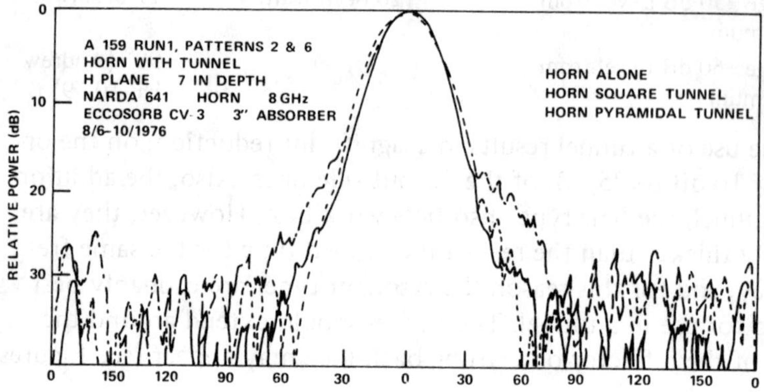

(b) H PLANE

Figure 6.5 Patterns of horn with tunnels – 7 in. depth

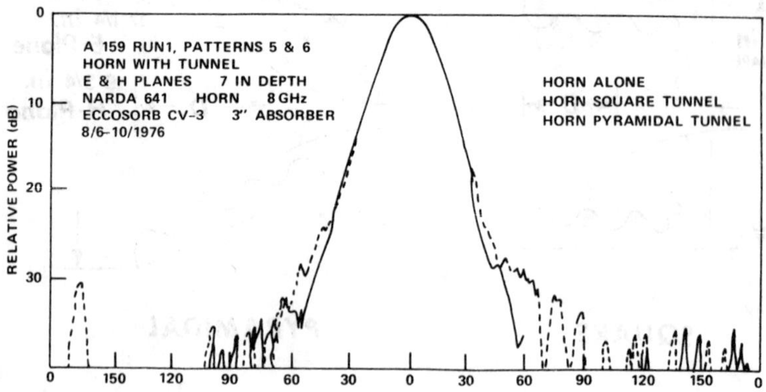

Figure 6.6 E and H plane pattern with square tunnel

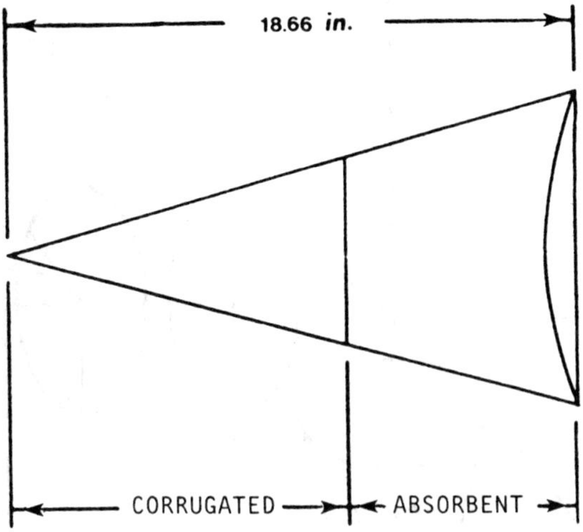

Figure 6.7 Lens system

Corrugated horn millimeter wave antennas with absorber tunnels have been developed for communications systems. The antennas had half-power beamwidths of about two degrees at 37 GHz. It was reported that sidelobes beyond 30 degrees off axis were more than 60 dB down and sidelobes beyond 80 degrees were more than 75 dB down. This configuration was also used as a feed for plano-hyperbolic lens systems, as shown in Figure 6.7, and for off-set-fed parabolic reflectors. It was reported that the best performance was obtained from the off-set paraboloid with an absorber lined shroud. This shroud was a tunnel, or cylinder, as shown in Figure 6.8.

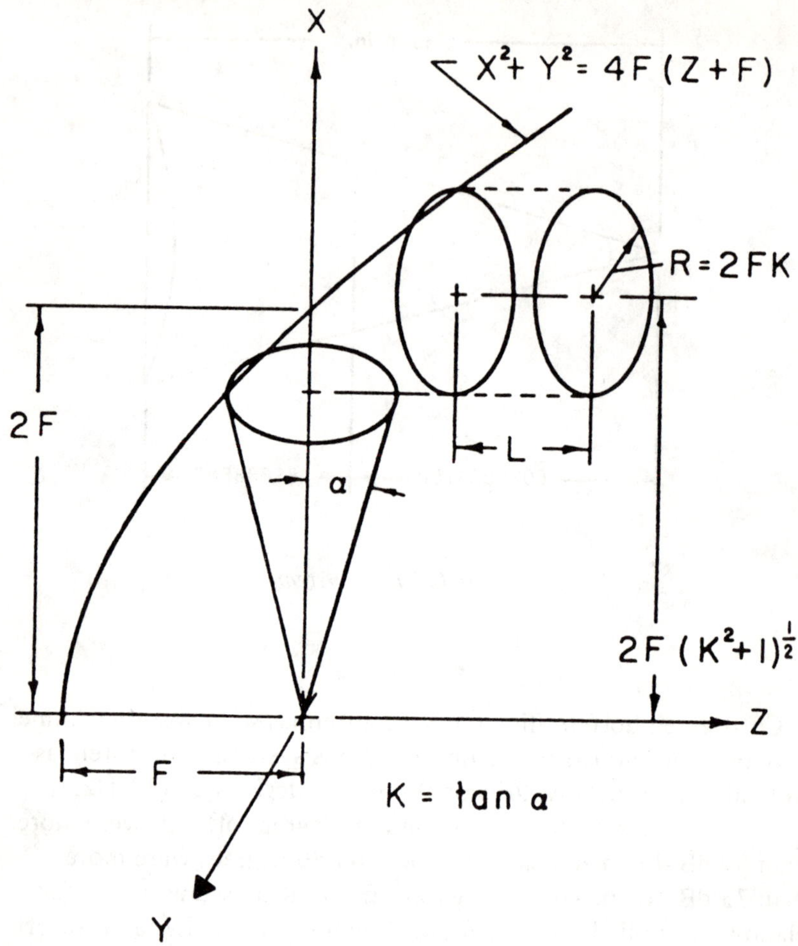

Figure 6.8 Geometry of offset-fed paraboloidal reflector

REFERENCES

[1] Shaoys, J., *Investigation of the Tunnel Antenna,* June 1958, AD 152444.

[2] Ryan, C.E., *A Geometrical Theory of Diffraction Analysis of the Radar Cross Section of a Sectionally Continuous Second-Degree Surface of Revolution,* Tech. Report 2430-4, Electro Science Laboratory of the Ohio State University.

[3] Silver, S., *Microwave Antenna Theory and Design.* New York: McGraw-Hill, 1949.

[4] Blacksmith, P., Jr., *Method of Reducing Far Out Side Lobes,* AFCRC TR 57115, ASTIA Document AD133707, Tech Report.

[5] Andrews Corporation, Catalog 25, *High Performance Shielded Antennas.*

[6] *Final Report on Theoretical and Experimental Investigations on Methods of Reducing Antenna Sidelobes,* Melpar, AD 152394.

[7] "Experimental Investigations of Antenna Side Lobe Reduction Using Tunnel Techniques," Melpar, Jan 1957, AD 117055.

[8] "Theoretical and Experimental Investigations On Methods of Reducing Antenna Sidelobes," AFCRC Contract No. AF 19 (604)-1898 Melpar, Inc.

[9] "Tests of Microwave Absorbing Materials for Tunnel Antennas," Melpar, Inc., May 1957.

[10] Yu, J.S., "Diffraction By Conducting Walls of Finite Thickness," AD 623077.

[11] Green, R.B., "The General Theory of Antenna Scattering," The Antenna Laboratory, Ohio State University, AD 429186.

[12] Ryan, C.E., "A Geometrical Theory of Diffraction Analysis of the Polar Cross-Section of a Sectionally-Continuous Second-Degree Surface of Revolution," Tech. Report 2430-4 Electro-Science Laboratory, Ohio State University, March 1968.

[13] Dybdal, R.B., "Horn Antenna Sidelobe Reduction Using Absorber Tunnels," IEEE 1977 International Symposium on Antennas and Propagation.

[14] Chung, G., and N. Yoshi Yuki, "Sidelobe Control of Aperture Antenna by Dielectric Rim Loading," IEEE 1975 International Symposium on Antennas and Propagation.

Figures 6.2 and 6.3 Figure courtesy of the Andrew Corporation, 10500 W. 153 Street, Orlando Park, Indiana.

7

HORN, REFLECTOR ANTENNAS

This section briefly describes several versions of horn reflector antennas. Basically a horn reflector antenna consists of a pyramidal horn illuminating a section of a parabolic reflector.

HORN, REFLECTOR COMBINATIONS

The Parabolic Horn Reflector

This is a pyramidal horn feeding a section of parabolic reflector. The main advantage of this configuration is that the horn acts as an electromagnetic shield which reduces the far-out sidelobes. The parabolic reflector changes the spherical horn wavefront into a plane wavefront.

After appropriate simplifications have been made, the characteristic equations [1] for the vertically polarized far-field radiation patterns become:

Half-Power Beamwidth$_\theta$ = 50.4 λ/d = wavelength
Null width$_\theta$ = 114.6 λ/d
Sidelobe$_\theta$ = −13.2 dB
Half-Power Beamwidth$_\phi$ = 68.9 λ°/a
Null width$_\phi$ = 172 λ°/a
Sidelobe$_\phi$ = −23 dB

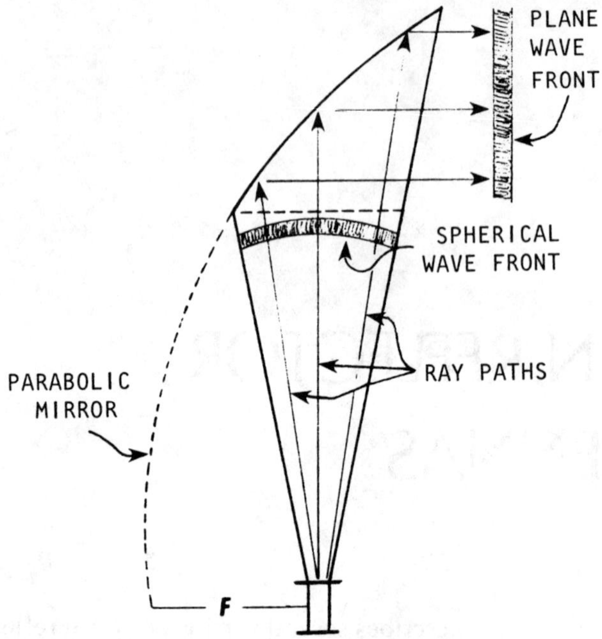

Figure 7.1 Shielded parabolic horn

Measurements were made [2] for a horn reflector with the following dimensions:

F = 14.898 inches
B_1 = 9.9 degrees
R_1 = 24.052 inches
R_2 = 34.252 inches

The following results were obtained (see Figure 7.3):

Half-Power Beamwidth	8.4 degrees
First Null Width	21.0 degrees
First Sidelobe Levels	–23.2 dB, –24.2 dB
Radiation Less Than	–30 dB from ± 16 degrees to ± 180 degrees
Radiation Less Than	–40 dB from ± 24 degrees to ± 180 degrees
Radiation Less Than	–50 dB from ± 41 degrees to ± 170 degrees
Backlobe Level	–45.3 dB

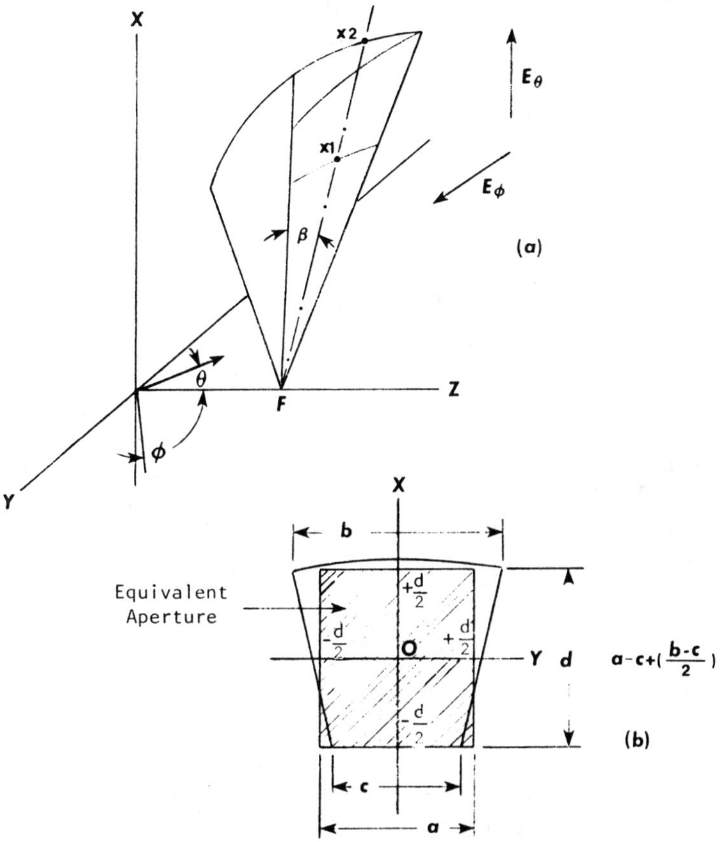

Figure 7.2 (a) Parabolic horn coordinate system; (b) equivalent aperture

This same antenna was evaluated at 34.625 GHz with the following results (see Figure 7.4):

	Vertical Polarization	Horizontal Polarization
Half-Power Beamwidth	2.4 degrees	1.7 degrees
First Sidelobe Levels	−20 dB, −23.7 dB	−11.5 dB, −12.3 dB
Radiation Less Than	−30 dB from ±5° to ± 180°	± 10° to ± 180°
Radiation Less Than	±40 dB from ± 7° to ± 180°	± 13° to ± 180°
Radiation Less Than	−50 dB from ± 16° to ± 180°	± 23° to ± 180°
Backlobe Level	−50 dB	>−55 dB

The point at which the −30 dB, −40 dB, and −50 dB levels start, for a given polarization, occur at about the same number of beamwidths from broadside for both frequencies.

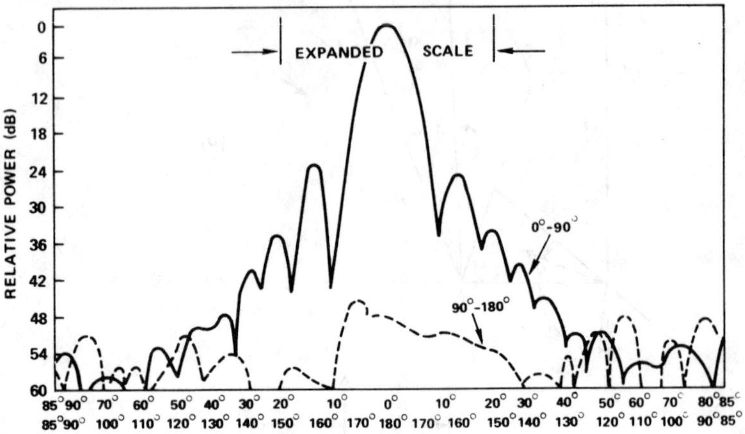

Figure 7.3 Vertically polarized azimuth pattern of parabolic horn at 9375 MHz

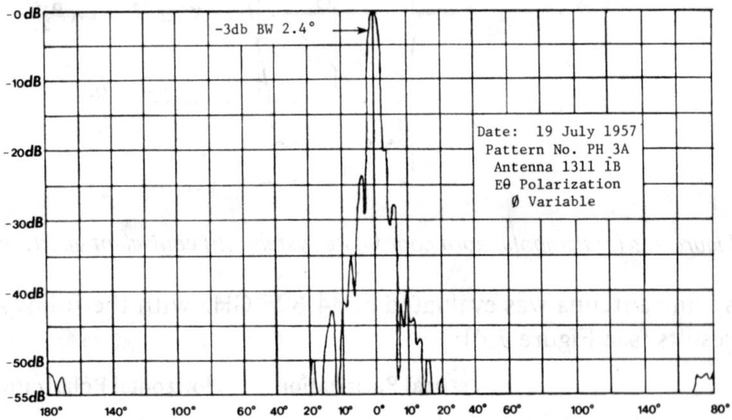

Figure 7.4 Azimuth plane radiation pattern of parabolic horn at 34.625 MHz

Parabolic Horn Reflector with a Correcting Lens

The previously described parabolic horn may be considerably shortened in height if a correcting lens is used. The microwave lens is placed in the mouth of the feed horn below the reflecting surface (see Figure 7.5).

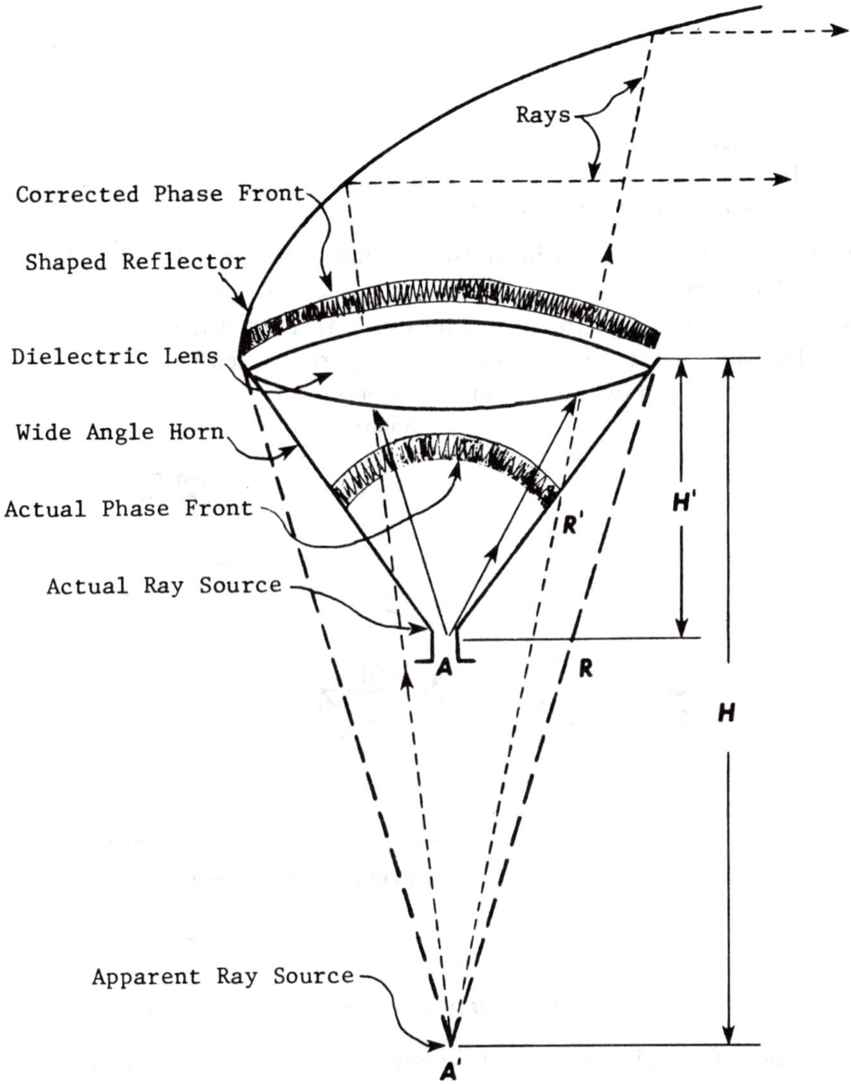

Figure 7.5 Corrective lens for wide angle parabolic horn

The curvature of the convex surface of a plano-convex lens in air is given by [3]:

$$R = \frac{(n - 1)\,L}{n \cos a - 1} \tag{7.1}$$

L = focal length
R = distance from source to lens surface
n = index of refraction of delay lens.

The focal length, L, must be chosen so that a source at A will appear as an apparent source at A′. Since the lens is not small compared to the focal length, a thick lens modification is necessary. The location of the nodal points h and h′, are defined by [4] $h - h' = t(n - 1)/n$, where t = lens thickness.

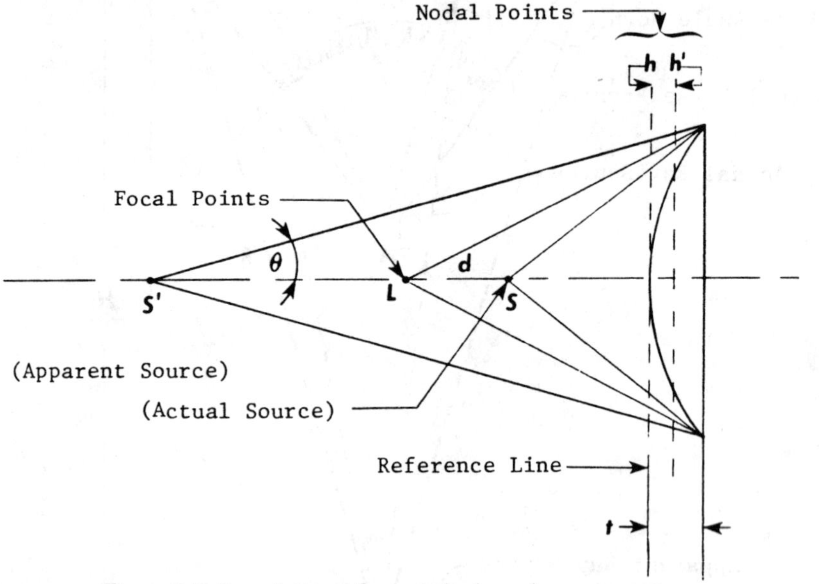

Figure 7.6 Description of parabolic horn lens parameters

By use of nodal points, the thin lens formula may be applied so long as the object distance S and the image distance S′ are measured respectively from h and h′ (see Figure 7.6).

The thin lens formula is given by [5]:

$$\frac{1}{S} + \frac{1}{S'} = \frac{1}{L} \tag{7.2}$$

The thin lens formula is given by [5]:

$$\frac{1}{S} + \frac{1}{S'} = \frac{1}{L} \tag{7.2}$$

The simulated flare angle θ may be realized by putting the source at S, in which case S′ becomes negative and the thin lens formula then reads:

$$\frac{1}{S} - \frac{1}{S'} = \frac{1}{S} \tag{7.3}$$

Whenever a wave traveling in one medium encounters a second medium of different density, some of the wave is reflected and the rest is transmitted into the second medium. Figure 7.7 indicates the percent transmitted and reflected as a function of the angle of incidence. The effective distribution at the radiating aperture is then the product of the aperture illumination of the feed horn times the tapered illumination of the lens. The lens may be used to lower the first sidelobes of the parabolic horn by selecting a lens with the proper illumination taper.

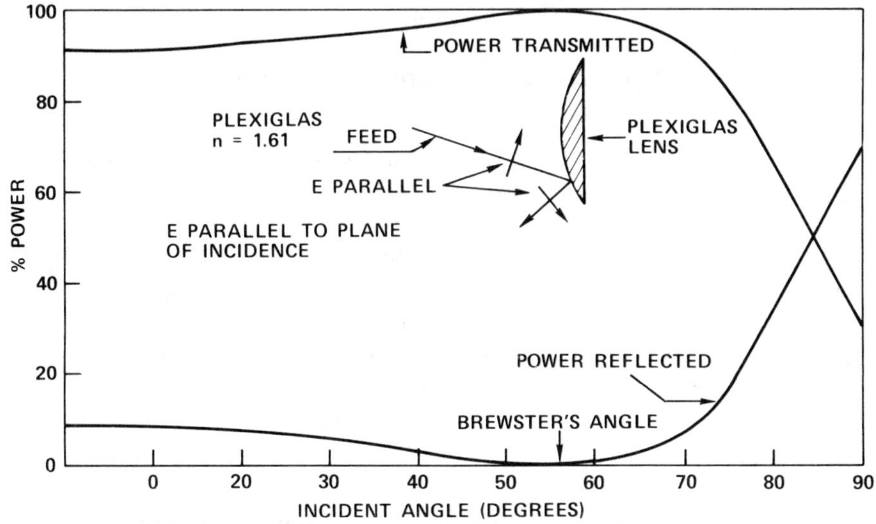

Figure 7.7 Percent power reflected and transmitted through plexiglas as a function of angle of incidence

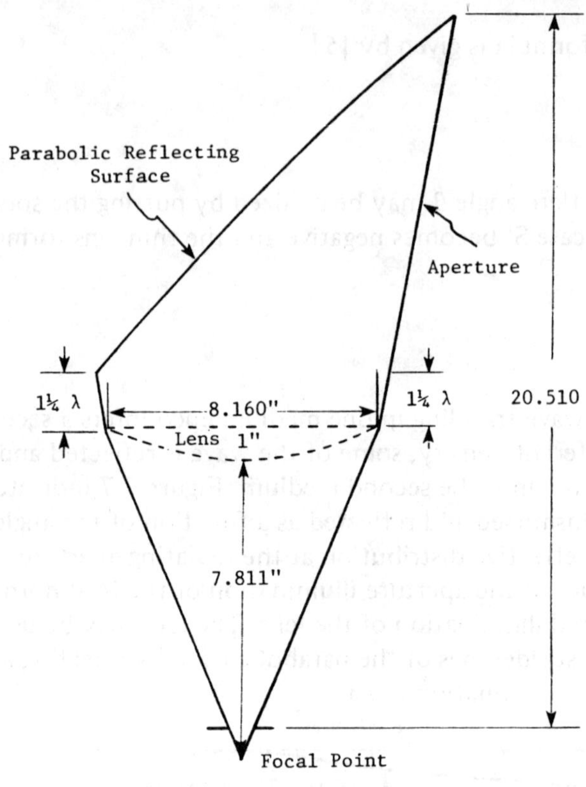

Figure 7.8 Parabolic horn with lens antenna

Figure 7.8 represents a parabolic horn with correcting lens for which the following results apply:

	Azimuth	Elevation
Half-Power Beamwidth	9.5°	9.5°
First Sidelobe Level	−28 dB	−17 dB
Radiation Less Than	−30 dB from ± 25° to ± 180°	± 52° to ± 180°
Radiation Less Than	−40 dB from 38° to 180°	97° to 180°
	−51° to −180°	−106° to 180°
Backlobe Level	−48 dB	− 48 dB

The parabolic horn with correcting lens is superior to the unmodified parabolic horn. It takes up less room and it has lower sidelobes. The level of the first sidelobes can be fixed, to some extent, by the taper of the correcting lens. Figures 7.9 and 7.10 are the azimuth and elevation patterns, respectively, of the parabolic horn with correcting lens.

The Shaped Horn Reflector

The shaped horn reflector antenna is the same as the previously described horn reflector antennas except that, instead of a parabolic reflector, the reflector is shaped to give an elevation pattern that conforms to the cosecant squared law. This was done for the case of vertical polarization. The disadvantage of this antenna is that the lens needed may be large.

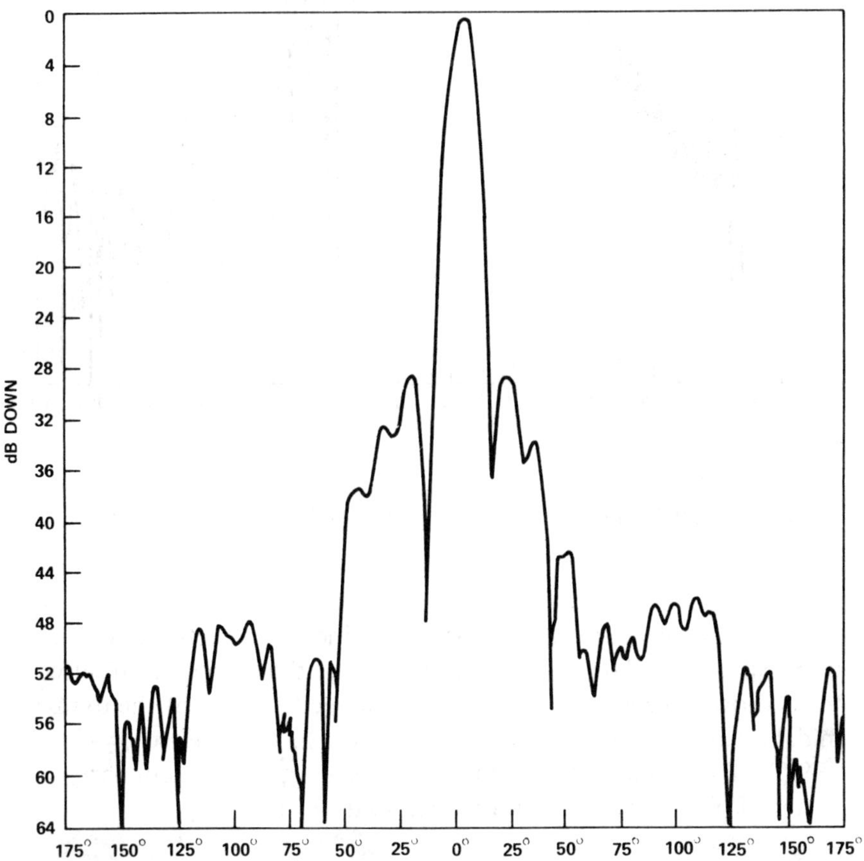

Figure 7.9 Azimuth pattern of parabolic horn with shaped lens antenna

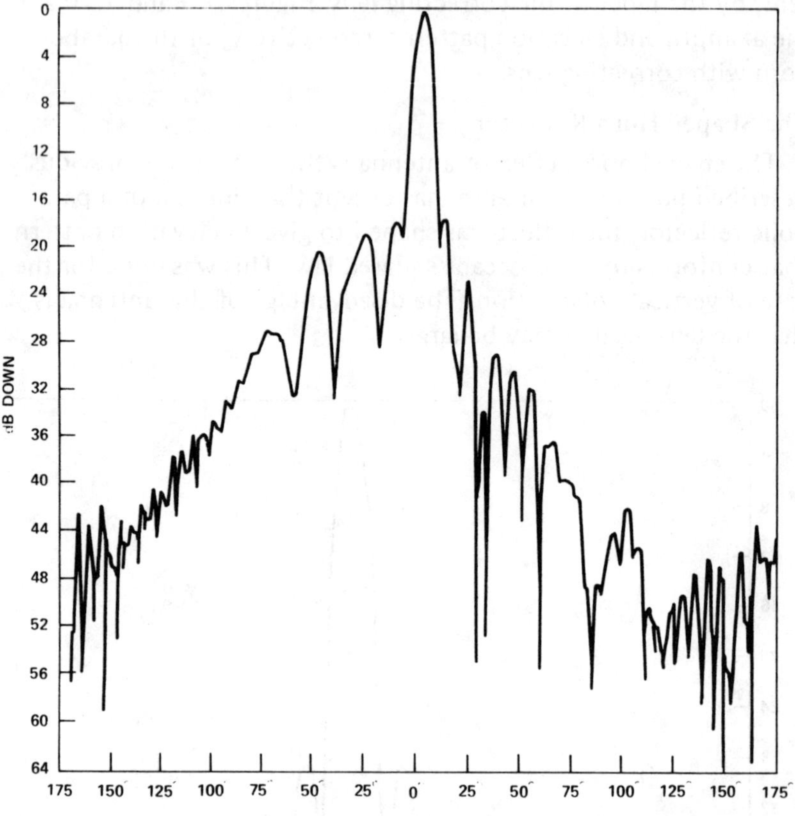

Figure 7.10 Elevation pattern of parabolic horn with lens antenna

The shape of the reflector was determined by a method described by S. Silver [1]. The pertinent geometry of the shaped horn reflector is shown in Figure 7.11. The E-plane feed pattern of the pyramidal horn is assumed to be uniform from $\psi_1 = -80.1°$ to $\psi_2 = -99.9°$. Then, since all of the power appearing in the secondary pattern $P(\theta)$ must appear in the feed pattern $I(\psi)$:

$$\int_{\psi_1}^{\psi_2} I(\psi)\,d\psi = K \int_{\theta_1}^{\theta_2} P(\theta)\,d\theta \qquad (7.4)$$

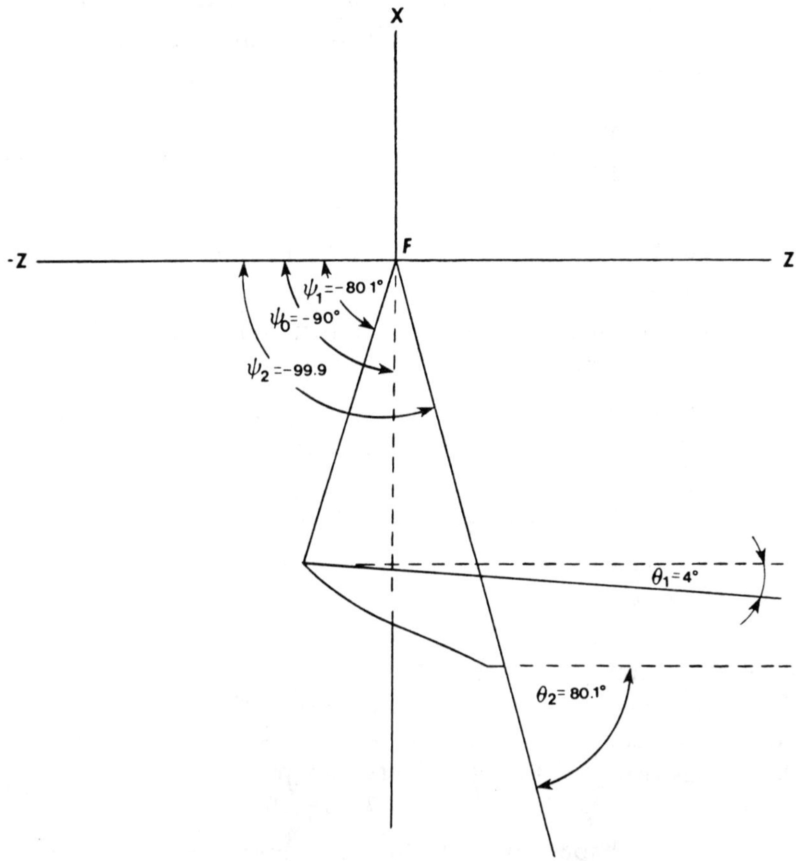

Figure 7.11 Shaped horn geometry

Sine I(ψ) is assumed to be constant from ψ_1 to ψ_2:

$$K = \frac{\psi_2 - \psi_1}{\int_{\theta_1}^{\theta_2} P(\theta)\, d\theta} \qquad (7.5)$$

When P(θ) is replaced by $\csc^2 \theta$, θ may be found from

$$\theta = \text{arc cot} \quad \cot\theta_1 + \frac{\cot\theta_2 - \cot\theta_1}{\psi_2 - \psi_1}(\psi - \psi_1) \qquad (7.6)$$

Arbitrary limits were set on θ_1 and θ_2 of 4 and 80.1 degrees respectively, then values of $\theta(\psi)$ are substituted in the general expression which determines the contour of the shaped reflector:

$$P(P_o) = \int_{\psi_1}^{\psi_2} \tan \frac{-\theta}{2} \, d\psi, \tag{7.7}$$

where p = distance from origin to a point on the reflector surface and $p = p_o$ at $\psi = \psi_1$.

A value of 25.16 inches was assigned to p_o to compare the shaped horn with the parabolic horn.

The geometrical optics techniques employed here are only valid for apertures of many wavelengths. The results obtained with this antenna [5] are:

	Azimuth
Half-Power Beamwidth	2.4°
First Sidelobe Levels	-30 dB
Second Sidelobe Levels	-37.2 dB
Third Sidelobe Levels	-41 dB, -45.2 dB
Radiation Less Than	-30 dB from ± 4.4° to ± 180°
Radiation Less Than	-40 dB from ± 7.6° to ± 180°
Radiation Less Than	-50 dB from +14° to 180° -17° to -180°

Patterns for the shaped horn reflector are shown in Figures 7.12 and 7.13.

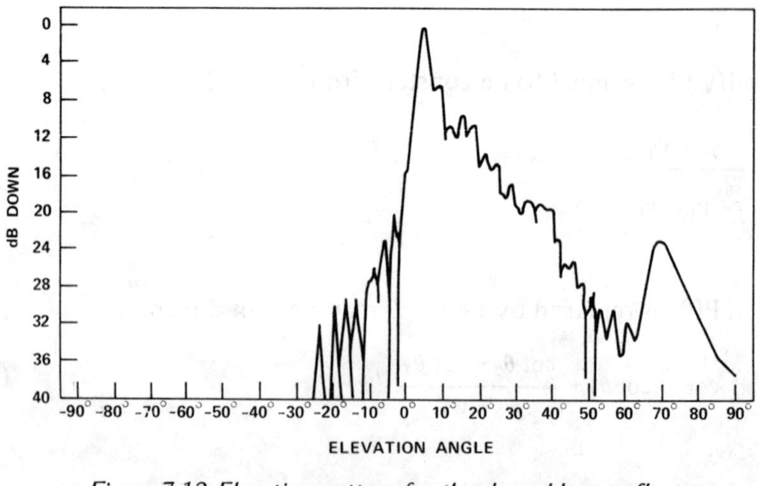

Figure 7.12 Elevation pattern for the shaped horn reflector

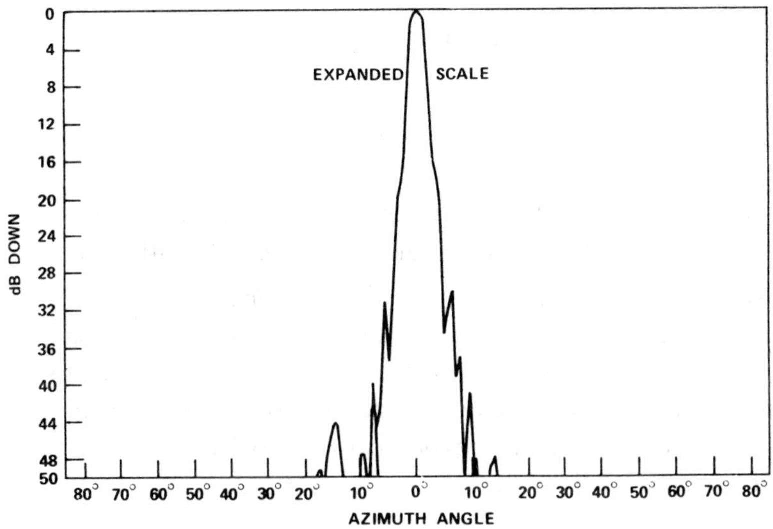

Figure 7.13 Azimuth pattern for the shaped horn reflector

The Casshorn Antenna

The Casshorn antenna was developed [7] to be a compact, low noise, high gain antenna. Curves will be presented below for the model developed in the above referenced study.

The Casshorn is a combination of the horn-reflector antenna with the Cassegrainian focusing technique. The horn-reflector concept is used because it results in very low sidelobes. However, the depth required for the horn is a disadvantage of this antenna. The Cassegrain antenna is compact because it uses a hyperboloidal reflector resulting in a virtual feed source that is much farther away than the actual feed. The Casshorn also uses an off-set feed which eliminates subreflector blocking. The overall antenna is, then, comprised of an off-set feed illuminating a hyperboloid which reflects the feed energy to the paraboloidal surface resulting in a plane wavefront. The antenna is shielded on three sides and is very compact.

Referring to Figure 7.14, the equation which describes the parabolic section is:

$$r = \frac{2f}{1 + \sin a}, \quad +a_o < a < -a_o \qquad f = \text{focal length} = OF \tag{7.8}$$

The parabolic surface is formed by rotating the parabolic section about the axis OX.

Using the focal length of the paraboloid f, and the horn flare angle a_o, the aperture height D can be defined as $D = 4f \tan a_o$.

Instead of having an actual source at F, a virtual source at F may be realized by putting the actual source at F′ and placing a concave hyperbolic reflecting surface between F′ and F. The hyperbolic surface makes the rays reflecting from it appear as though they had come from F. The hyperbola with foci at F and F′ is defined from Figure 7.15 as:

$$\epsilon - \rho\ 2a \tag{7.9}$$

also, $[\epsilon = (\rho \sin \psi)^2 + (2f - \rho \cos \psi)^2]^{1/2}$

which with Equation (1) yields:

$$\rho = \frac{f^2 - a^2}{a + f \cos \psi} \tag{7.10}$$

This equation represents a family of curves satisfying the focusing requirement. To define the surface of the design, "a," which is dependent on the maximum angle intercepted by the hyperbola, ψ_o, must be specified. In order to make the design compact, the hyperbola should intercept the parabolic section at the point determined by the flare angle a_o. From Figure 7.16, we may write:

$$\tan \psi_o = \frac{r_1 \sin a_o}{2f - r_1 \cos a_o}, \quad \text{where } r_1 = \frac{2f}{1 + \sin a_o}. \tag{7.11}$$

Then:

$$\psi_o = \tan^{-1} \frac{\sin a_o}{1 + \sin a_o - \cos a_o} \tag{7.12}$$

at $\psi = \psi_o$, Equation (1) becomes:

$$2a = r_1 - \rho_1 \tag{7.13}$$

which, from the geometry of Figure 7.16 can be written as

$$\rho_1 = \frac{r_1 \sin a_o}{\sin \psi_o} \tag{7.14}$$

Substituting Equation (7.15) and the expression for r_1 into (7.14) yields:

$$a = \frac{f(\sin \psi_o - \sin a_o)}{\sin \psi_o(1 + \sin a_o)} \tag{7.15}$$

The hyperbolic surface is obtained by rotating about the line FF′ the section defined by (2), (3), and (6). The complete design is specified in terms of the paraboloidal focal length, f, and the horn flare angle, a_o. The ratio a/f and the feed half-angle ψ_o are plotted as functions of a_o in Figure 7.17.

A symmetrical feed distribution at F′ results in an asymmetrical amplitude distribution over the aperture plane because of the difference in path length from F′ to the edges of the parabola. The space taper that results (from Figure 7.19) is:

$$T(\text{dB}) = 20 \log \frac{r_1}{r_2} \tag{7.16}$$

or in terms of a_o:

$$T(\text{dB}) = 20 \log \frac{1 - \sin a_o}{1 + \sin a_o} \tag{7.17}$$

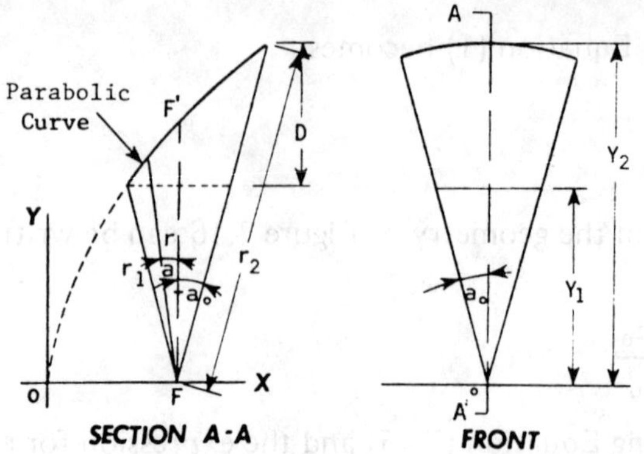

Figure 7.14 Horn-reflector geometry (side and front views)

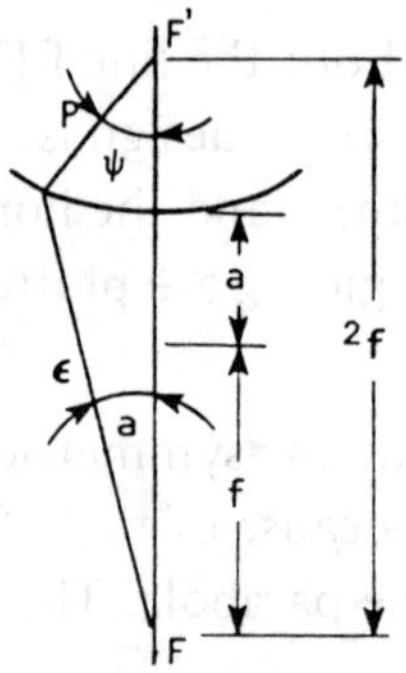

Figure 7.15 Hyperbola geometry

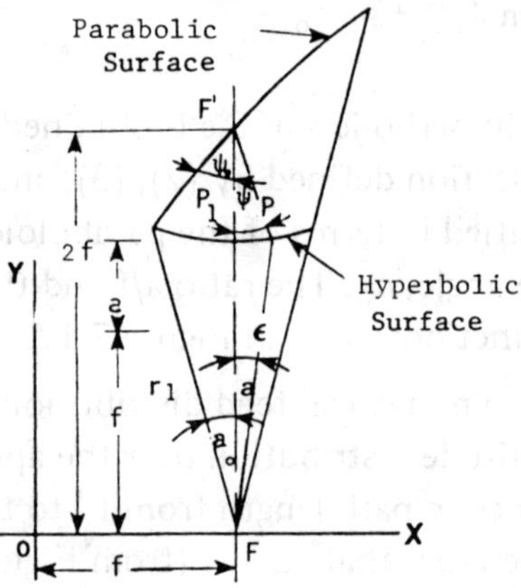

Figure 7.16 Evolution of Cassegrain horn-reflector

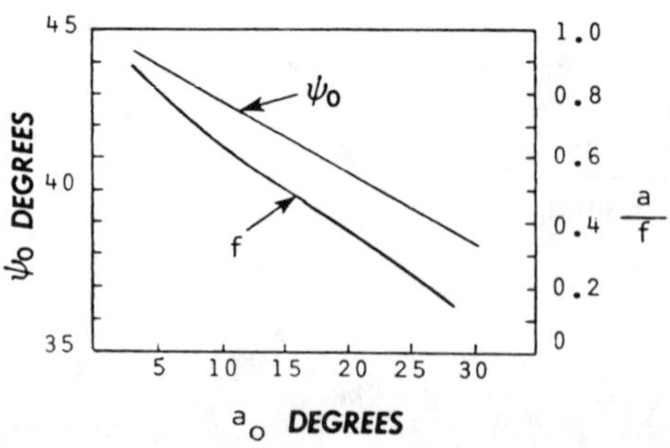

Figure 7.17 Design parameter relationships

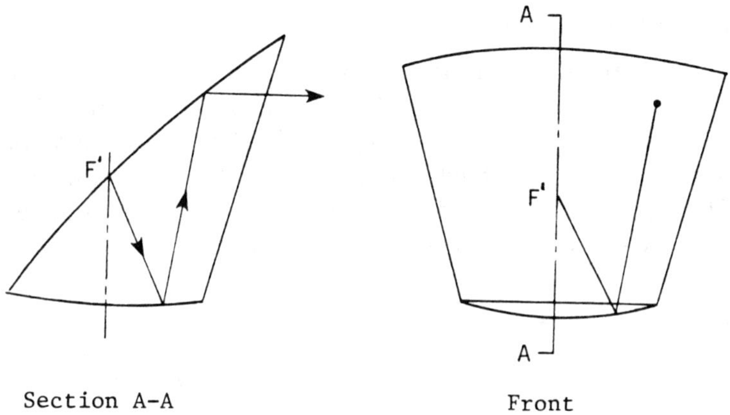

Figure 7.18 Typical ray (side and front view)

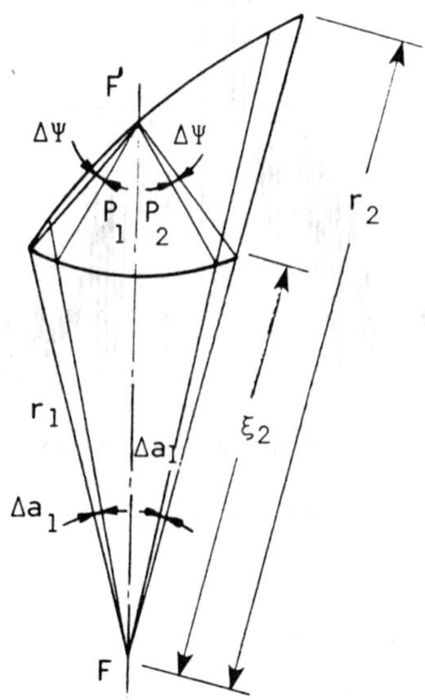

Figure 7.19 Geometry for space attenuation

Additional views of the antenna are presented in Figure 7.18, along with a typical ray path.

A small experimental model was fabricated [2] with the following dimensions: equivalent horn flare angle, $a_o = 15°$; parabolic focal length, f = 30 inches; feed angle $2\psi_o = 83°$ and an hyperbola constant, a = 0.48. The feed was an X-band circular horn with E- and H-plane 10 dB beamwidths of 65° and 80°. Representative patterns are shown in Figures 7.20 and 7.21.

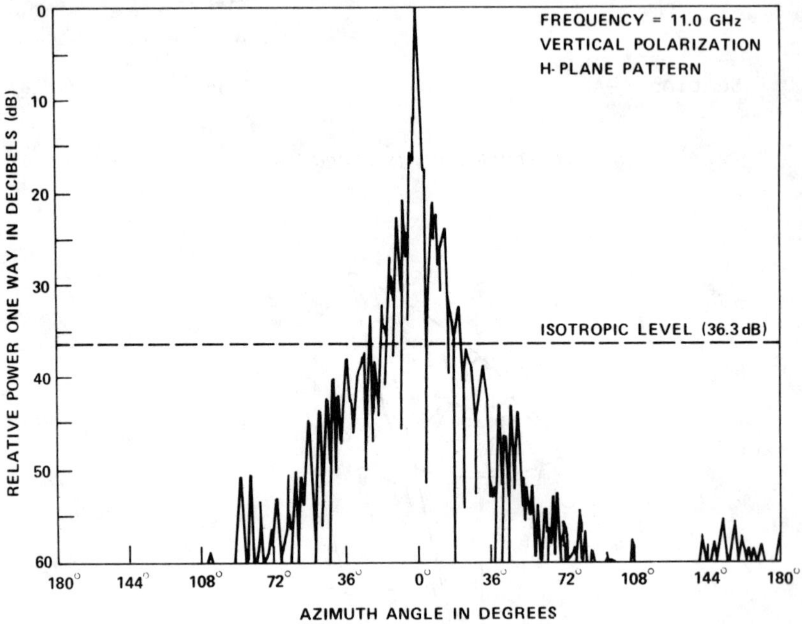

Figure 7.20 Horn reflector antenna with half blinder attached

	Azimuth	Elevation
Half-Power Beamwidth	4°	5°
First Sidelobe Levels	-21 dB	-21 dB
Radiation Less Than	-30 dB from ± 18° to 180°	+20° to +180° -10° to -180°
Radiation Less Than	-40 dB from ±36° ± 180°	± 45° to ± 180°
Radiation Less Than	-50 dB from ± 54° ± 180°	± 90° to ± 180°
Backlobe Level	56 dB	

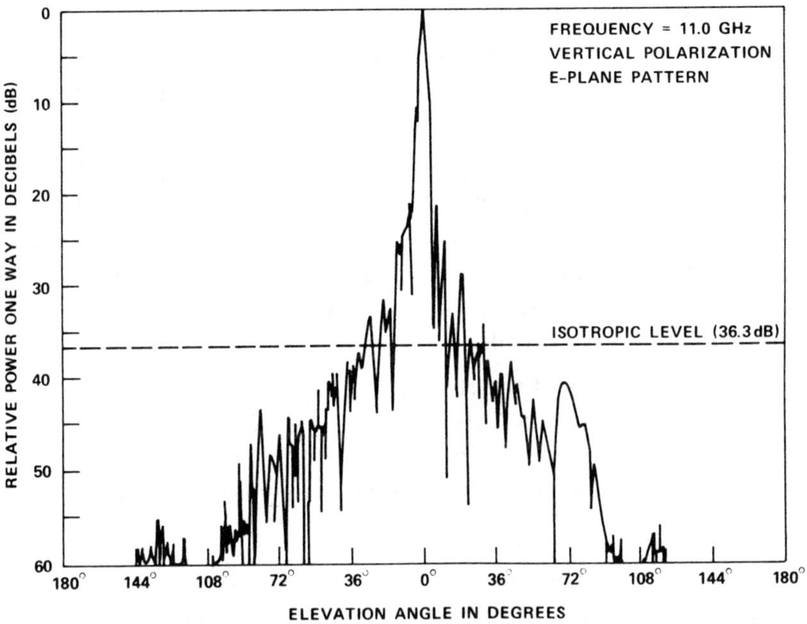

Figure 7.21 Elevation plane pattern

Measured gain at 11 GHz was 36.3 dB, corresponding to an aperture efficiency of 50 percent for a dish 32 inches in diameter. It was believed that this figure could have been improved had the model been put together with greater care.

The Casshorn antenna has characteristics similar to the parabolic horn antenna but is much smaller. Also, it eliminates the aperture blocking problem common to most Cassegrain antennas by using the off-set feed configuration.

Conclusions

In this section the pertinent design equations and representative sidelobe levels of some horn reflector antennas have been presented.

The following table summarizes these results.

The basic horn reflector antenna may be shortened by using the correcting lens. The first sidelobes and the backlobe of the horn reflector with correcting lens are lower than the corresponding lobes of the basic horn reflector.

The shaped horn reflector antenna exhibits an asymmetrical, steep skirted elevation pattern which is useful for locating targets with respect to elevation.

The Casshorn antenna is the most compact of the horn-reflector combinations. It combines the Cassegrain antenna feed technique (use of a subreflector) with the shielding properties of the horn-reflector antenna. The radiation characteristics are similar to the basic horn-reflector antenna. Aperture blocking is avoided through utilization of an offset feed.

The sidelobe levels of the referenced Casshorn could have been lowered by improved fabrication techniques. As a whole, however, the horn reflector antennas do not significantly reduce the sidelobe levels compared to an unmodified Cassegrain antenna.

Use of Blinders for Reducing Sidelobes

The Bell System has developed a half-blinder (Figure 7.20) for use with horn reflector antennas. Its small size makes it easy to fabricate and the wind loading is relatively negligible. Half blinders were initially developed to reduce the 90° off-axis sidelobe from -54 dB to -62 dB. This was accomplished and at least -62 dB was maintained for all sidelobes from 35° to 180°.

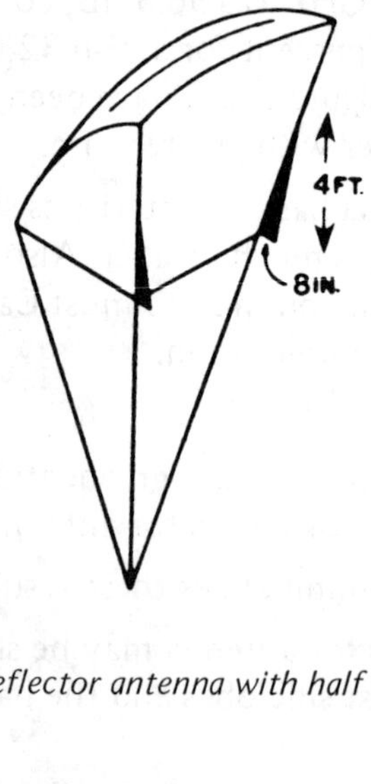

Figure 7.22 Horn reflector antenna with half blinder attached

REFERENCES

[1] S. Silver, op. cit.

[2] "Final Report on Theoretical and Experimental Investigations on Methods of Reducing Antenna Sidelobes," Melpar, AD 152 394.

[3] Kraus, J.D., *Antennas,* New York: McGraw-Hill, 1950.

[4] Robertson, R.S., *Introduction to Physical Optics.* New York: D. Van Nostrand, 1929.

[5] Jenkins, F.A. and H.E. White, *Fundamental of Optics.* New York: McGraw-Hill, 1929.

[6] Final Report on Theoretical and Experimental Investigations on Methods of Reducing Antenna Sidelobes, Melpar, AD 152 394.

[7] S.R. Jones and K.S. Kelleher, "A New Low Noise, High Gain Antenna," *IEEE International Convention Record,* Part 1, 1963.

[8] Thomas, D.T., "A Half Blinder for Reducing Certain Sidelobes in Large Horn Reflector Antennas," Correspondence, *IEEE Transactions on Antennas and Propagation,* May 1971.

Figures 7.3 and 7.4 Final Report on Theoretical and Experimental Investigations on Methods of Reducing Antenna Sidelobes, Melpar, AD 152 394.

Figure 7.20 Thomas, D.T., "A Half Blinder for Reducing Certain Sidelobes in Large Horn Reflector Antennas," Correspondence, *IEEE Trans. Antennas Propag.,* May 1971.

8

ARRAYS

In this section the sidelobe levels and frequency dependent characteristics of arrays will be considered.

USE OF ARRAYS FOR SIDELOBE SUPPRESSION

The total field at a distant point can be represented by

$$E = 1 + e^{j\psi} + e^{2j\psi} + e^{3j\psi} + \ldots + e^{j(n-1)\psi} \tag{8.1}$$

where ψ is the total phase difference between adjacent sources and is

$$\psi = d_i \cos\phi + \delta$$

where δ is the phase difference of adjacent sources, and

$$d_i = \frac{2\pi d}{\lambda}, \text{ spacing in radians.}$$

The sum E of the above infinite series may be determined as follows:

Multiplying by $e^{j\psi}$ gives:

$$E\, e^{j\psi} = e^{j\psi} + e^{j2\psi} + e^{j3\psi} + \ldots. + e^{jn\psi}. \tag{8.2}$$

By subtraction

$$E - E\, e^{j\psi} = 1 - e^{jn\psi} \tag{8.3}$$

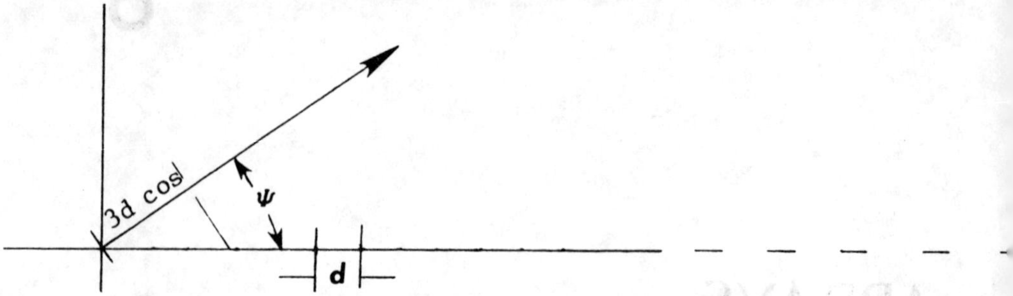

Figure 8.1 Linear Arrays of n Isotropic Point Sources of Equal Amplitude and Spacing.

then dividing by $1 - e^{j\psi}$ gives

$$E = \frac{1 - e^{jn\psi}}{1 - e^{j\psi}} . \qquad (8.4)$$

Equation 8.4 may be rewritten as

$$E = \frac{e^{j(n\psi/2)}}{e^{j(\psi/2)}} \quad \frac{e^{j(n\psi/2)} - e^{-j(n\psi/2)}}{e^{j(\psi/2)} - e^{j(\psi/2)}} \qquad (8.5)$$

or

$$E = e^{j\psi(n-1/2)} \frac{\sin |n\psi/2|}{\sin |\psi/2|} .$$

When the phase of the field is referred to the center of the array, then

$$E = \frac{\sin (n\psi/2)}{\sin (\psi/2)} \qquad (8.6)$$

When $\psi = 0$, Equation 8.6 becomes indeterminate, but E may be determined as the limit of (8.6) as ψ approaches zero: $E = n$. This is the maximum value that E can attain and therefore the normalized expression for E becomes:

$$E = \frac{1}{n} \frac{\sin (n\psi/2)}{\sin (\psi/2)} \qquad (8.7)$$

The field as expressed by (8.7) is sometimes known as the "array factor" [1].

The far-field pattern is a maximum when $\psi = 0$, that is, the fields from the sources all arrive at a distant point in phase when $\psi = 0$. When the array is a linear array of isotropic sources of equal amplitude and phase it is known as a uniformly illuminated "broadside array". The overall pattern from an array may be found by pattern multiplication if the array is composed of similar radiators. Essentially, this is multiplying the field pattern of an individual source by the field pattern of an array of isotropic sources. Therefore, the total field pattern of an array may be found by multiplying the individual source pattern times the "array factor".

Since ψ depends on the spacing, d, of the elements, which is usually expressed in wavelengths and is commonly one-half of a wavelength, most arrays are designed for a single frequency. Frequency independent arrays may be made up of equiangular or log-periodic elements; we will look at them later.

For linear arrays of large numbers of isotropic point sources of equal amplitude and spacing, the amplitudes of the first few sidelobes are given, to a good approximation, by $E_{SL} \simeq 2/(2K+1)\pi$ where K is the sidelobe number. For instance, the first (K=1) sidelobe is 21 percent of the main beam. It can be seen from this equation that the sidelobe level doesn't depend on the number of sources.

The indicated first sidelobe level is therefore the best that can be expected from a uniformly illuminated linear array.

BINOMIAL DISTRIBUTION

However, with proper adjustment of the amplitudes of the sources, it is theoretically possible to eliminate sidelobes entirely. By way of comparison, the beamwidths and first sidelobe levels of several different amplitude distributions will be given for a five element array with half-wavelength spacing. Uniform distribution yields the maximum directivity (beamwidth of 23°) but the first sidelobes are 24 percent of the main beam. If however, the sources have amplitudes proportional to the coefficients of a binomial series, there are no minor lobes. This has been achieved at the expense of increasing the beamwidth to 31°.

OPTIMUM DISTRIBUTION

There is another distribution possible which results in a compromise between beamwidth and sidelobe level. This is known as the Dolph-Tchebyscheff [2] distribution and has the properties that if the beamwidth is specified, the resulting sidelobe levels will be the smallest possible. Conversely, if the maximum sidelobe level is specified, the resulting beamwidth will be the narrowest possible. For the same five element array, if the sidelobe level is specified as being 20 dB below the main beam, the resulting beamwidth is 27°.

If the uniformly illuminated array had source amplitudes equal to unity, then the amplitudes of the binomial distribution would be 1, 4, 6, 4, 1 and the amplitudes of the Dolph-Tchebyscheff distribution would be 1, 1.6, 1.9, 1.6, 1. The sidelobe levels are a function of the amplitude distribution over the aperture: the steeper the taper, the less the sidelobes as seen in the binomial distribution. However, the distribution has a broader beamwidth than the uniform distribution which has the narrowest beamwidth but the highest sidelobes. Evidently, an abrupt discontinuity in the distribution results in large sidelobes, while a gradually tapered distribution approaching zero minimizes the sidelobes. This is also the objective of placing a tunnel on a reflecting antenna or extending the dish in the form of a "deep dish" design. If there is no spillover past the main reflector, in a Cassegrain antenna for example, and no illumination of the edge of the reflector, then there should be no radiation in the far-out and backlobe region.

There are two major problems with simple arrays — for a given size, geometrical optics techniques provide higher gains and directivities than simple arrays and simple arrays are not broadband.

INCREASED DIRECTIVITY

The directivity may be increased by placing a large reflecting sheet behind the array. The sheet confines the radiation to the forward direction and increases the gain in this direction. Another way to increase the directivity is to use a broadside-area array. Here the radiators are arranged in a rectangular or square shape in a plane normal to the direction of maximum response. The principle of pattern multiplication is applicable here, for if a sharp pattern is multiplied by one nearly as sharp then the resulting beam is appreciably sharper than either one alone. For example, if $E_x(\theta,\phi)$, $E_y(\theta,\phi)$ and $E_z(\theta,\phi)$ are field patterns of the component linear arrays in the three coordinate directions and if it is desired to find the field in the x-direction, then the field in the y-z plane is $E_{yz} = E_y(\theta,\phi)\,E_z(\theta,\phi)$. For maximum directivity with minimum dimensions, $E_y(\theta,\phi)$ should equal $E_z(\theta,\phi)$.

FREQUENCY-INDEPENDENT ARRAYS

There are basically two kinds of antennas that could be said to be frequency-independent. The first kind evolved from the idea that the properties of an antenna, pattern and impedance, are determined by its shape and size specified in wavelengths. If, by arbitrary scaling, the transformed antenna is equal to the original one, then it is said to be frequency-independent. An antenna of this type satisfies the angle condition, which means its shape can be specified entirely by angles, not dimensions. Conical antennas composed of infinite cones of arbitrary cross section having a common apex, and equiangular antennas, which have their surfaces generated by equiangular spirals having a common axis and the same defining parameter, are two antennas satisfying the angle condition.

The second kind of frequency-independent antenna is derived from the idea that if an antenna becomes equal to itself by a particular scaling of its dimensions, like a ratio τ, it will have the same properties at a frequency f and at τf. If, for example, the impedance is a periodic function of log τ, the antenna is called a log-periodic. With τ close to 1, the variation of impedance with frequency will be small and the antenna is essentially frequency-independent.

The pattern of an array of periodic elements may be considered as the simple superposition of the patterns of the individual elements. For an array to be frequency-independent, the locations of the elements need to be defined by angles rather than distances. This implies that all of the elements have a common feed point. For a log-periodic array, although all of the element patterns are identical, they point in different directions. Because all elements appear to radiate from a common phase center, the elements are arranged in a spherical shape for a two-dimensional array or in a circular shape for a one-dimensional array.

The disadvantage of the two kinds of frequency-independent elements mentioned here is that they have small gains. For a two-element log-periodic array, for example, the E-plane half-power beamwidth is about 63° and the gain over a standard dipole is about 7.5 dB. Considering the various disadvantages of arrays, it would appear that the most effective approach at this time would be to use parabolic antennas designed for low interference with other systems.

ARRAYS WITH NON-UNIFORMLY-SPACED ELEMENTS

Finally, considerable interest has been generated in nonuniformly-spaced arrays in which the elements have different spacings as they proceed from the center to the edge of the array. One of the anticipated benefits of this approach, as compared to uniformly spaced elements, is the reduction in the number of elements required. This would be the same as broadening the bandwidth and the scanning range of the array for a given number of elements.

The other reason for interest in nonuniformly or space-tapered arrays was that it was felt that the space-tapered equivalent of the excitation-tapered Dolph-Tchebyscheff could be made.

This would simplify the design of arrays, since space-tapering is much simpler than excitation-tapering. However, when the spacing is small between elements, considerable mutual coupling can result. A survey conducted [3] of space-tapered arrays concluded that there was no way of determining the optimum placement of the elements. However, various approximation methods that are adequate for most applications have been proposed [4]; the reader is referred to the references for additional information on the subject.

LARGE PHASED ARRAYS

The ever-increasing requirements for antenna resolution, gain, and sidelobe suppression have led to steerable reflector antennas on the order of as large as 200-300 feet in diameter. When such high performance is required, the large phased array becomes an attractive alternative for economic reasons. The interferometer spacing in an array is limited by the requirement to suppress interferometer (grating lobe) effects. The requirement for narrower and narrower beamwidths, therefore, leads to a requirement for very large numbers of elements.

There is a trade-off, therefore, between the practical number of elements and achievable resolution and sidelobe levels. Experience has shown that practical arrays of randomly spaced elements can be designed to have narrow beam, reasonably low sidelobe levels and wide bandwidth with a limited number of array elements. Lo developed the theory for such arrays of randomly spaced elements. The theory provides a probabilistic prediction for performance using basic parameters. This prediction can be used to determine constraints prior to detailed calculations or design. Arrays with randomly spaced elements have the following properties [43]:

1. The number of elements required depends primarily on the desired sidelobe level — this number is, in general, much less than the corresponding number required for uniform element spacings.
2. The array resolution depends principally on the aperture dimension.
3. The array directive gain is proportional to the number of elements used.
4. When the number of elements is fixed, the resolution and bandwidth corresponding to an aperture can be improved by a factor of 10, 100 or more using random element spacing, without substantial risk of having a higher sidelobe level.

One method for arriving at a figure-of-merit for large array designs is to assess the number of elements required vs. the resulting resolution and sidelobe level. The random element spacing array theory assumes that most elements are widely spaced (to avoid mutual coupling effects) and that the number of elements involved is fairly large (on the order of 50 or more). Lo [44] derived the distribution of the sidelobe level for a linear array and got

$$\Pr(\text{sidelobe level below } r) = (1 - e^{-Nr^2})^{[4a]}$$

where Pr(•) denotes "probability of";

[4a] denotes the larger of the two integers nearest 4a;

N = total number of array elements;

a = aperture dimension in wavelengths;

r = the sidelobe level with the main beam main beam maximum normalized to unity.

The above equation, plotted as a family of universal curves is

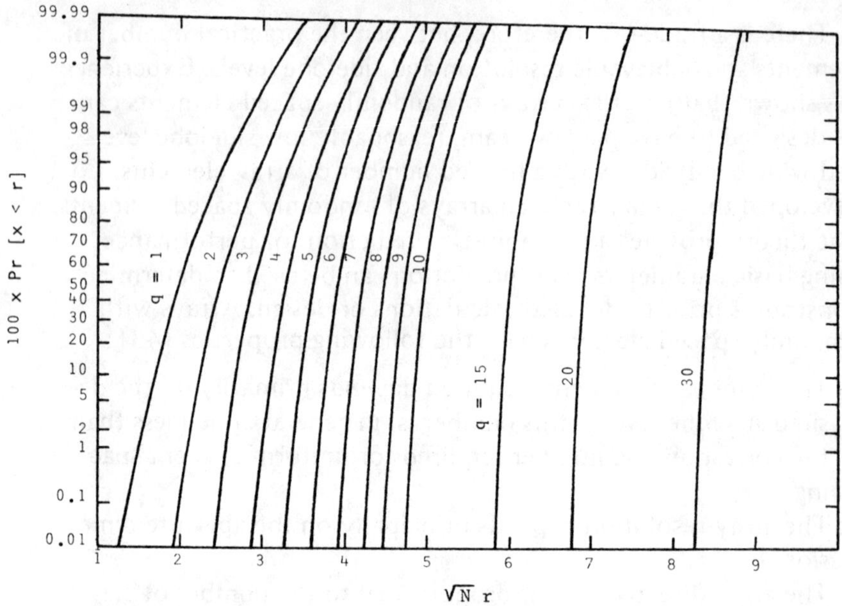

Figure 8.2 Cumulative probability that sidelobe level ⩽ r versus $\sqrt{N}r$.

contained in Figure 8.2. The curves provide the probability that the sidelobe level is below r vs. r $\sqrt{N}$ for various aperture dimension indexes, q, where

$$q = \log_{10}(a) .$$

This approach can be applied to a two dimensional array (a x b) by replacing q with (p + q) where

$$a = 10^q \text{ and } b = 10^P,$$

and modifying the cumulative probability scale as follows. For a given probability level (Pr_1) in Figure 8.2 (for a linear array) find a second probability level (Pr_2) for the two-dimensional array, by finding a value of x that satisfies

$$Pr_1 = 1 - e^{(-x^2)}$$

then find Pr_2,

$$Pr_2 = \text{erf}(x)$$

where

$$\text{erf}(x) \triangleq \frac{2}{\sqrt{\pi}} \int_0^x e^{-t^2} dt.$$

Conclusions

The linear array composed of n isotropic sources with equal amplitude and spacing has the smallest beamwidth but the highest sidelobes of the considered arrays. Theoretically, sidelobes can be entirely eliminated when a binomial amplitude distribution is used to feed the array elements. This, however, results in a broader beamwidth than that of the uniformly excited array.

The best compromise between the two above mentioned distributions is known as the Dolph-Tchebyscheff distribution. This feed amplitude distribution results in the lowest sidelobe level possible for a given beamwidth and, conversely, the narrowest beamwidth for a specified sidelobe level.

In general, frequency-independent arrays would not be applicable for shipboard use because of their low gains and broad beamwidths. However, frequency dependent arrays have one advantage over other type antennas. Very accurate aperture control is possible because of the individual elements that make up the array. Each element has its own feed and electronic beam steering is possible by controlling the feed amplitudes and phasing of the appropriate elements. Electronic beam steering is not possible with the geometric optics antennas such as the Cassegrain antenna and hence must be physically steered.

Generally speaking, arrays can be made that have gains comparable to reflector type antennas. Beamwidths less than 2° are possible with modest size arrays. However, although arrays can give performance comparable to reflector type antennas, the complexity of arrays will probably make them more expensive than parabolic reflector antennas.

REFERENCES

[1] Kraus, J.D., *Antennas.* New York: McGraw-Hill Book Company, 1950.

[2] Dolph, C.L., "A Current Distribution for Broadside Arrays Which Optimizes the Relationship Between Beamwidth and Sidelobe Level," *Proc. IRE,* Vol. 34, June 1946.

[3] Lo, Y.T., and S.W. Lee, "A Study of Space-Tapered Arrays," *IEEE Trans.,* Vol. AP-14, Jan. 1966.

[4] Schuman, H.K., and B.J. Strait, "On the Design of Unequally Spaced Arrays with Nearly Equal Sidelobes," *IEEE Trans.,* Vol. AP-16, July 1968.

[5] General Information for Radar Set, AN/SPS-48(V)(u), Vol. 1.

[6] Schuman, H.K., and B.J. Strait, "Computer Design of Unequally Spaced Arrays," *Sci. Rept.,* AFCRL-67-0552, Oct. 1967.

[7] Steifel, E.L., "Numerical Methods of Chebyshev Approximation," in *Proc. of a Symp. on Numerical Approximation,* R.E. Langer, Ed. Madison: University of Wisconsin Press, 1956.

[8] Tang, C.H., "On a Design Method for Nonuniformly Spaced Arrays," *IEEE Trans.,* Vol. AP-13, July 1965.

[9] "An Approximate Method of Designing Non-Uniformly Spaced Arrays," *IEEE Trans. on Antennas and Propagation (Communications),* Vol. AP-13, January 1965, pp. 177-179.

[10] Tang, C.H., and V.W. Chang, "Directive Gains of Nonuniformly Spaced Arrays," *IEEE Trans. on Antennas and Propagation (Communications),* Vol. AP-14, July 1966, pp. 505-507.

[11] Unz, H., "Linear Arrays with Arbitrarily Distributed Elements," *IRE Trans. on Antennas and Propagation,* Vol. AP-8, March 1960, pp. 222-223.

[12] King, D.D., R.F. Packard, and R.K. Thomas, "Unequally-Spaced, Broad-Band Antenna Arrays," *IRE Trans. on Antennas and Propagation,* Vol. AP-8, July 1960, pp. 380-385.

[13] Lo, Y.T., "A Non-Uniform Linear Array System for the Radio Telescope at the University of Illinois," 1960 Spring URSI and IRE Joint Meeting in Washington, D.C. Subsequently published in the *IRE Trans. on Antennas and Propagation,* Vol. AP-9, January 1961, pp. 9-16, under the title, "The University of Illinois Radio Telescope," by G.W. Swenson and Y.T. Lo.

[14] Maffett, A.L., "Array Factors with Non-Uniform Spacing Parameters," *IRE Trans. on Antennas and Propagation,* Vol. AP-10, March 1962, pp. 131-136.

[15] Andreasen, M.G., "Linear Arrays with Variable Interelement Spacings," *IRE Trans. on Antennas and Propagation,* Vol. AP-10, March 1962, pp. 137-143.

[16] Ishimaru, A., "Theory of Unequally-Spaced Arrays," *IRE Trans. on Antennas and Propagation,* Vol. AP-10, November, 1962, pp. 691-702.

[17] Yen, J.L., and Y.L. Chow, "On Large Non-Uniformly Spaced Arrays," *Canad. J. Physics,* Vol. 41, January 1963, p. 1.

[18] Ishimaru, A., and Y.S. Chen, "Thinning and Broadbanding Antenna Arrays by Unequal Spacings," *IEEE Trans. on Antennas and Propagation,* Vol. AP-3, January 1965, pp. 34-42.

[19] Bernard, G., and A. Ishimaru, *Tables on the Anger and Lommel-Weber Functions.* Seattle: University of Washington Press, 1962.

[20] Chow, Y.L., and J.L. Yen, "Grating Plateaux of Planar Non-Uniformly Spaced Arrays," presented at the 1964 Fall URSI Meeting, Urbana, Ill.

[21] Skolnik, M.I., G. Nemhauser, and J.W. Sherman, III, "Dynamic Programming Applied to Unequally-Spaced Arrays," *IEEE Trans. on Antennas and Propagation,* Vol. AP-12, January 1964, pp. 35-43.

[22] Lo, Y.T., *On the Theory of Randonly Spaced Antenna Arrays,* Antenna Lab., Dept. of Elec. Eng., Univ. of Illinois, Urbana, Tech. Rept. 1 under NSF G 14894, 1962. Part of this report appeared in "A Mathematical Theory of Antenna Arrays with Randomly Spaced Elements," *IEEE Trans. on Antennas and Propagation,* Vol. AP-12, May 1964, pp. 257-268.

[23] Lo. Y.T., "A Probabilistic Approach to the Problem of Large Antenna Arrays," *Radio Science,* Vol. 68D, September 1964, pp. 1011-1019.

[24] Harrington, R.F., "Sidelobe Reduction by Non-Uniform Element Spacing," *IRE Trans. on Antennas and Propagation,* Vol. AP-9, March 1961, pp. 187-192.

[25] Baklanov, Y.V., V.L. Pokrovshi, and G.I. Surdotovich, "A Theory of Linear Antennas with Unequal Spacing," *Radio Eng. and Electronic Phys.,* no. 6, June 1962, pp. 905-913.

[26] Rabinowitz, S.J., and R.F. Kolar, "Statistical Design of Space-Tapered Arrays," presented at the 1962 12th Annual Symposium on USAF Antenna Research and Development Program, Univ. of Illinois, Urbana.

[27] Skolnik, M.I., J.W. Sherman III, and F.C. Ogg, Jr., "Statistically Designed Density-Tapered Arrays," *IEEE Trans. on Antennas and Propagation,* Vol. AP-12, July 1964, pp. 408-417.

[28] Maher, T.M., and D.K. Cheng, "Random Removal of Radiators from Large Linear Arrays," *IEEE Trans. on Antennas and Propagation,* Vol. AP-11, March 1963, pp. 106-111.

[29] Lo, Y.T., "Sidelobe Level in Non-Uniformly Spaced Antenna Arrays," *IEEE Trans. on Antennas and Propagation (Communications),* Vol. AP-11, July 1963, p. 511.

[30] "High Resolution Antenna Arrays with Elements at Quantized Random Spacings," presented at the 1964 International Conference on Microwaves, Circuit Theory, and Information Theory, Tokyo.

[31] Cover, T.M., "Antenna Pattern Distributions from Random Arrays," Rand Corp., Santa Monica, CA. Memo. RM-3402-PR, January 1963.

[32] Gardner, M., "Mathematical Games," *Sci. American,* Vol. 210, March 1964, pp. 120-128.

[33] Ogg, F.C., "Steerable Array Radar," *IRE Trans. on Military Electronics,* Vol. MIL-5, April 1961, pp. 80-94.

[34] Lo, Y.T., and S.W. Lee, *A Study of Elliptical and Related Arrays.* Antenna Lab., Elec. Eng. Dept., Univ. of Illinois, Urbana Tech. Rept. 7 under contract NEL 30508A, October 1964.

[35] Chow, Y.L., "On Grating Plateaux of Non-Uniformly Spaced Arrays," *IEEE Trans. on Antennas and Propagation,* Vol. AP-13, March 1965, pp. 208-215.

[36] Willey, R.E., "Space Tapering of Linear and Planar Arrays," *IRE Trans. on Antennas and Propagation,* Vol. AP-10, July 1962, pp. 369-377.

[37] Hyneman, R.F., "A Technique for the Synthesis of Line-Source Antenna Patterns Having Specified Sidelobe Behavior," *IEEE Trans.,* Vol. AP-16, July 1968.

[38] Taylor, T.T., "Design of Line-Source Antennas for Narrow Beamwidth and Low Sidelobes," *IRE Trans.,* Vol. AP-3, Jan. 1955.

[39] Hyneman, R.F., and R.M. Johnson, "A Technique for the Synthesis of Shaped-Beam Radiation Patterns with Approximately Equal-Percentage Ripple," *IRE Trans.,* Vol. AP-15, Nov. 1967.

[40] Unz, H., "New Methods for Synthesis of Non-Uniformly Spaced Antenna Arrays," AD 405866.

[41] Justice, R., "Sidelobe Suppression by Pattern Multiplication," *IRE Trans. AP,* April 1956.

[42] Jen, L., "Suppression of Sidelobes of Linear Arrays," Emmanuel College, Mass., Nov. 1965, AD 631063.

[43] R.A. Monzingo and T.W. Miller, *Introduction to Adaptive Arrays,* New York: John Wiley & Sons, 1980.

Figure 8.2 Jen, L., "Suppression of Sidelobes of Linear Arrays," Emmanuel College, Mass., Nov. 1965, AD 631063.

9

MODIFICATION of PILLBOX ANTENNAS for LOW SIDELOBE LEVELS

The pillbox antenna is an effective antenna for producing a line source because of the relative simplicity of the feed as compared to an array antenna.

Many pillboxes have feeds which are in the aperture in order to illuminate the parabolic reflector at the back. Two parallel plates are connected by the parabolic-cylinder reflector to make an enclosed structure. The parallel plates act as a waveguide. Since the feed is in the middle of the aperture, some aperture blocking of the pattern is inevitable. Aperture blocking by the feed results not only in a shadow region but also in undesired reflections because of the finite size of the feed. These reflections have been reduced somewhat [1] by placing absorbing material on the outer surface of the feed. With such an application, deeper nulls and more symmetrical side lobes approaching those predicted were obtained.

The gain of a pillbox antenna is approximately 60 per cent of the gain of a uniformly illuminated aperture of equivalent size. The sidelobe levels of a simple pillbox can be as low as 24 dB, but only over a very narrow frequency band where the back radiation from the feed and the sidelobes from the aperture cancel. With a few per cent change in frequency the sidelobes may increase to 14 dB down from the main beam.

Research [2] indicates that the sidelobe levels for a folded pillbox can be reduced to below the 30 dB level. This is achieved by multiple feeding of the pillbox. The basic idea is to superimpose several simple distributions to obtain an overall distribution that

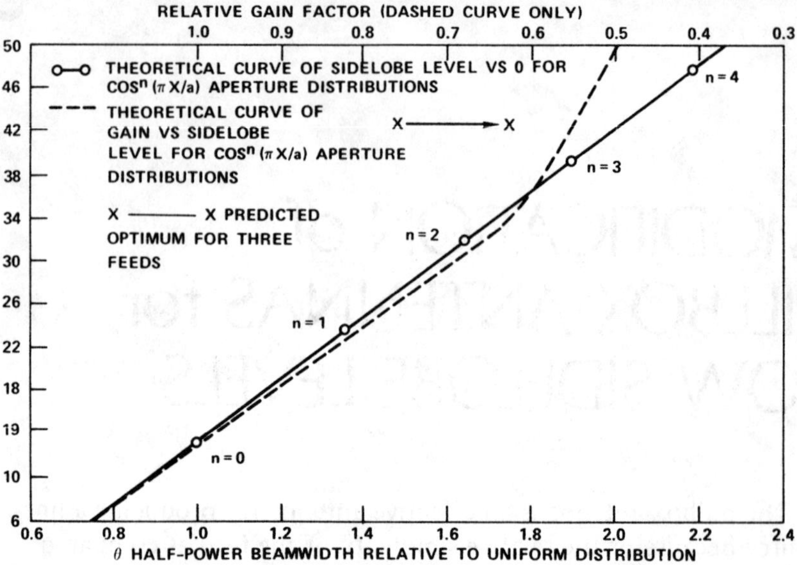

Figure 9.1 Sidelobe level vs. half-power beamwidth for $cos^n(-X/a)$ aperture distributions.

has the desired characteristics. For instance, the uniform distribution has the highest gain of the in-phase distributions and a narrow beamwidth but it has fairly high sidelobes (−13 dB). The more the distribution is tapered toward the edge of the aperture, the smaller the sidelobes will be. Cosine distributions, for example, of orders one, two, and three have sidelobe levels down 23 dB, 32 dB, and 40 dB, respectively, from the main beam. Along with the higher orders, however, go broader beamwidths and lower gains (Figure 9.1).

Theoretically, a distribution made by superimposing several simple distributions can have a much better distribution. The radiation pattern of a cosine-squared distribution added to the pattern of a uniform distribution results in a pattern with very low sidelobes. This is because, as seen in Figure 9.2A, the zeros and the maxima of the two distributions coincide and the maxima of the sidelobes are of opposite signs. If the amplitude of the uniform distribution is about 7 per cent of the maxima of the cosine-squared distribution, then the main lobes will add and the sidelobes will tend to cancel. This combination could result in sidelobes that are 43 dB below the main lobe. Also, the overall distribution will have an aperture efficiency that is greater than that of the cosine-squared distribution.

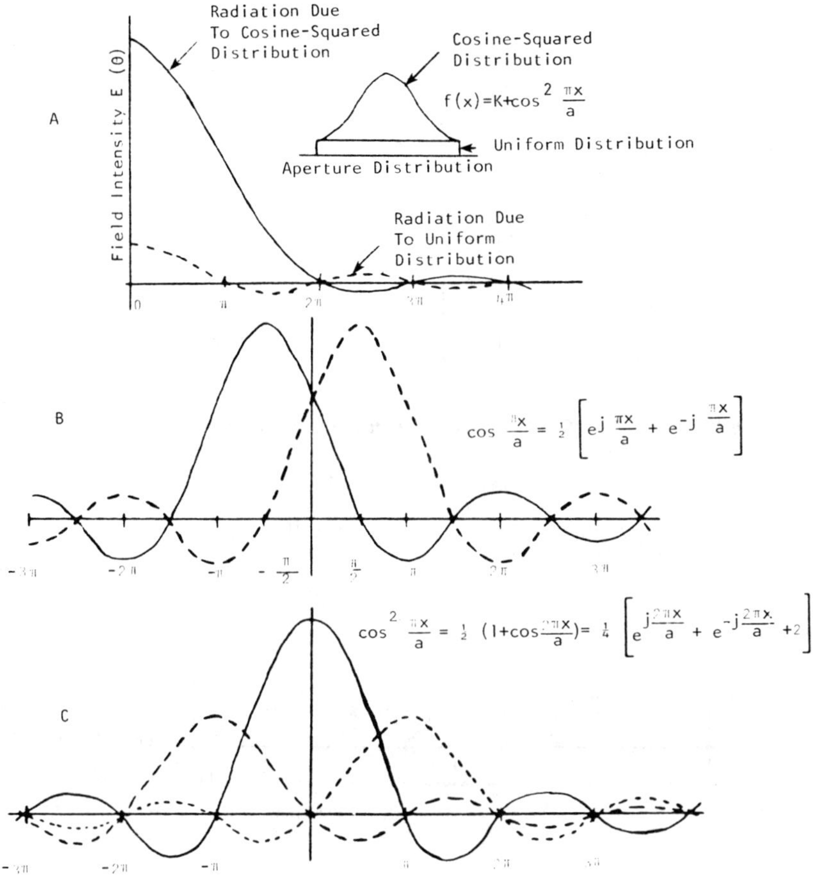

Figure 9.2 Radiation pattern components of some $\cos^n(\pi x/a)$ aperture distributions

A simple cosine function can be represented by two exponential terms as in Figure 9.2B. Each exponential term can be thought of as a uniform plane wave with a linear phase distribution. Since they are 180° out of phase, their sidelobes tend to cancel. Using the same approach, the cosine-squared distribution can be broken up into three terms. The radiation patterns for the three plane waves are shown in Figure 9.2C.

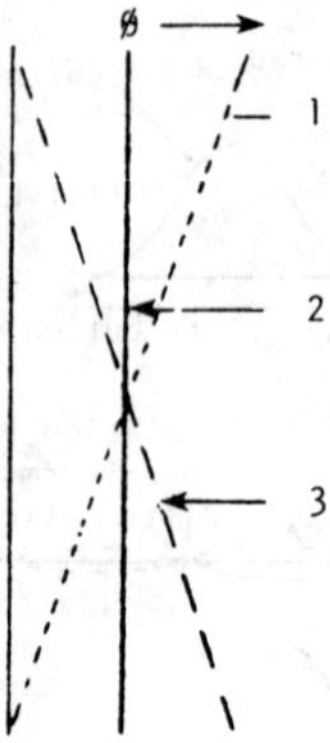

Figure 9.3 Component phase distributions across the aperture

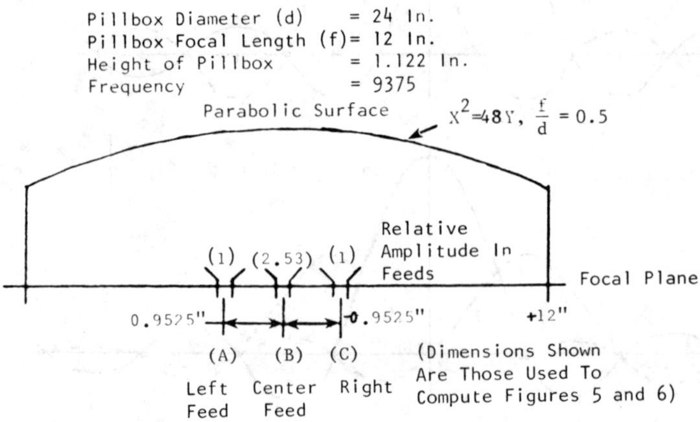

Figure 9.4 Method of superimposing distributions by multiple-feeding of the reflector

An analysis was made of the three-feed method to determine how closely, in theory, the combined cosine-squared and uniform distribution could be realized. Several factors tend to alter the actual distribution from ideal: (1) the directivity of each of the three feeds, (2) the power decrease at the edges of the pillbox due to increased path length, (3) the phase and amplitude distortion due to the offset feeds, (4) the unknown nature of the transmission around the bend in the pillbox, and (5) the effect of the side walls of the pillbox. The first three factors were accounted for in the analysis, and a reasonable assumption made about the fourth. The fifth factor could be determined experimentally or through diffraction theory.

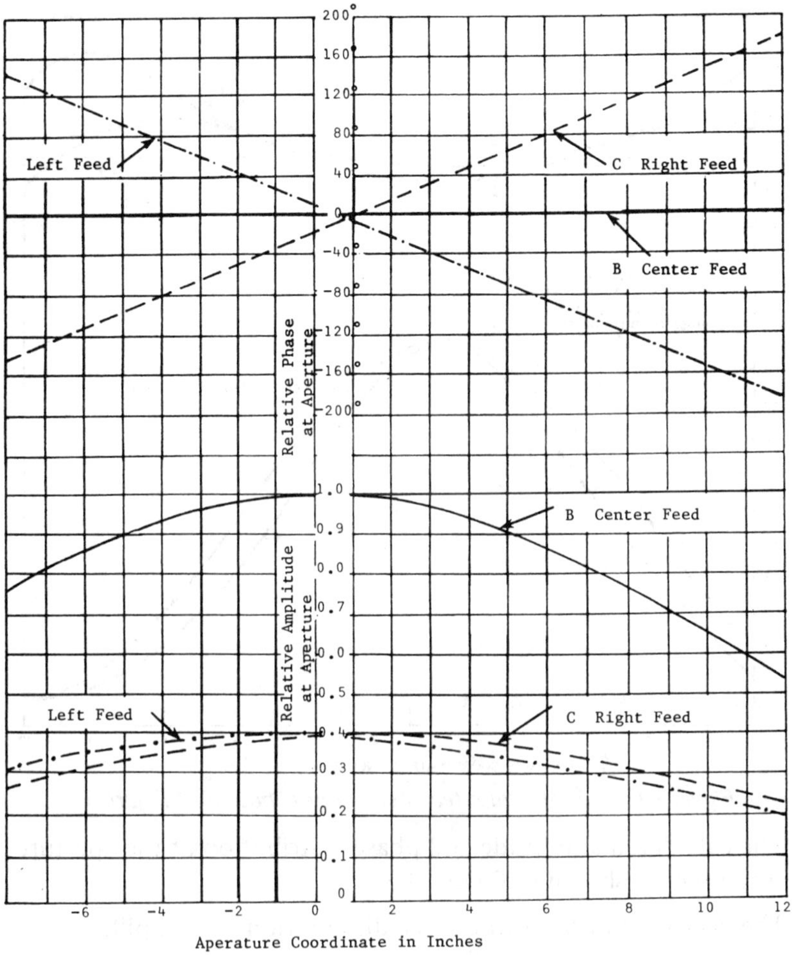

Figure 9.5 Computed aperture distribution of individual feeds

The initial analysis was performed for a horizontally polarized pillbox with the following dimensions: focal length (f) over diameter (d) of pillbox = 0.5, f = 12 inches, d = 24 inches, height of pillbox = 1.122 inches. The pillbox was operated at 9375 MHz. The ratio of the three in-phase amplitudes was assumed to be such that the amplitude at the aperture edge was 7 per cent of the maximum. This was required for optimum sidelobe cancellation. The outside feeds were offset so that the slope of their phase fronts was $2\pi/a$. Simple ray theory was used to determine the phase front.

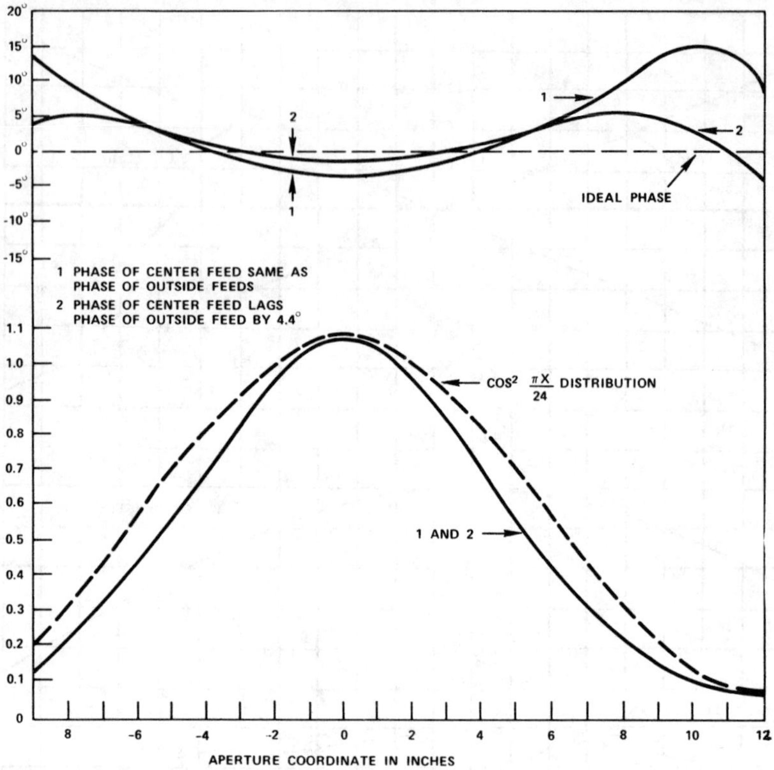

Figure 9.6 Sum of computed aperture distribution of Figure 5

The calculated amplitude and phase distribution at the aperture for each feed is shown in Figure 9.5.

The vector addition of the three distributions at the pillbox aperture is shown in Figure 9.6.

As can be seen, both the amplitude and phase distributions deviate somewhat from the ideal, even with the assumption of a perfect bend and no reflections from the side-walls. The phase front can be improved somewhat by changing the relative phase between the center and outside feeds (as seen from the curve of the 4.4 degree delay of the center feed).

It was originally decided to measure the field distributions at the mouth of the aperture and relate them to the far-field patterns. This, it was felt, would eliminate many of the variables entering into the secondary radiation pattern (such as external reflections). In this manner, the effect of small changes in the feed

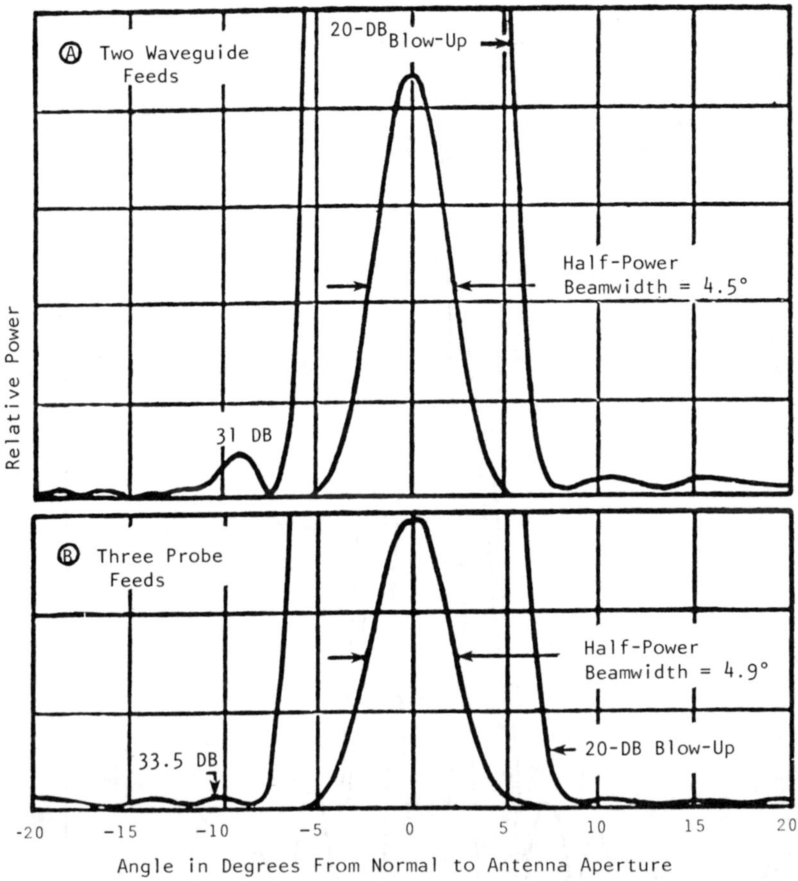

Figure 9.7 Patterns showing best sidelobe levels achieved with two and three probe feeds

structure would be more readily identified. However, because of such significant discontinuities as diffraction from the edges of the aperture and reflections from the side-walls, a smooth aperture distribution did not exist. Consequently, a large amount of computation would have been required to determine the far-field pattern and this approach was dropped.

Far-field patterns were taken for the case of two waveguide feeds located $\zeta = \pm 0.4$ inch off the centerline and are shown in Figure 9.7A. The 4.5 degree beamwidth and the −31 dB sidelobe level compare favorably with the 4.4 degree beamwidth and the −32 dB sidelobe level predicted. The asymmetry presented by the edges of the focal plane reflector is also apparent in Figure 9.7A.

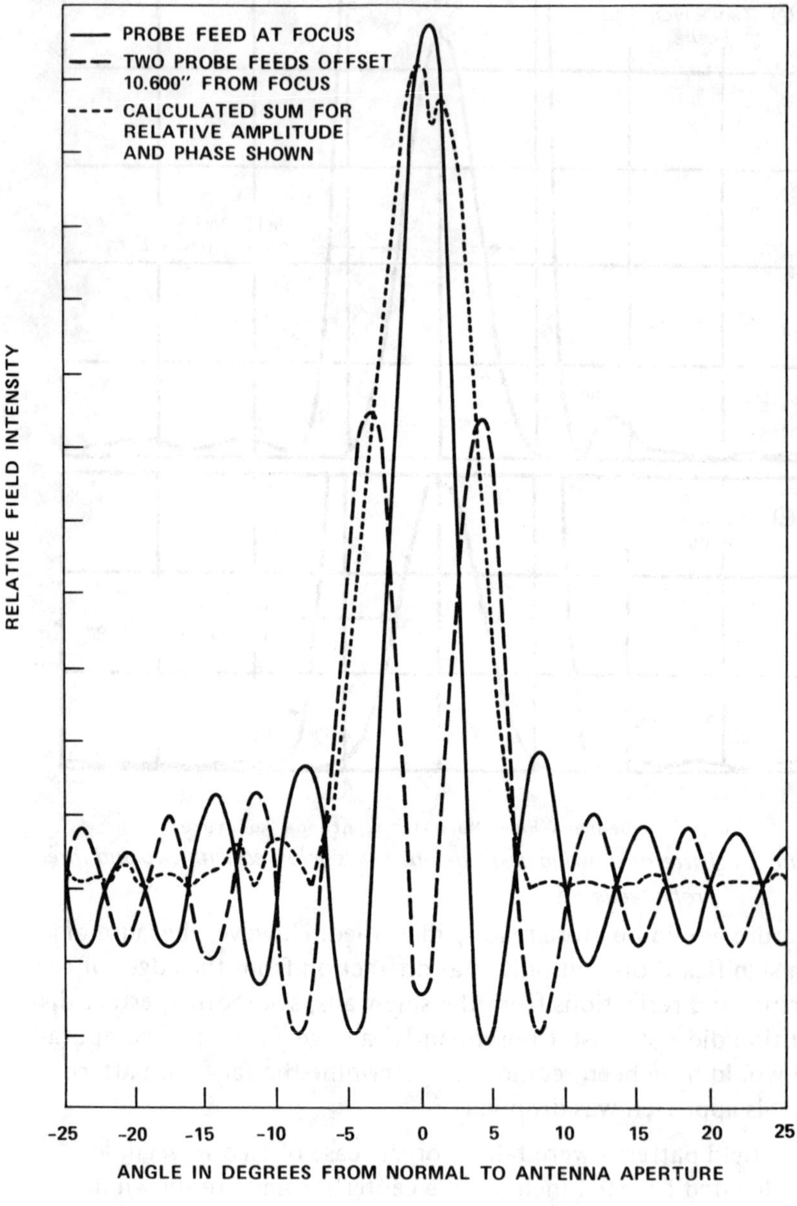

Figure 9.8 Measured patterns of pillbox with single probe feed and two superimposed offset probe feeds

For three probe feeds, the actual sidelobe level was down 33.5 dB with θ = 1.83 as compared to the theoretically best value of 43 dB with θ = 1.5, as seen from Figure 9.7B and Figure 9.1 (θ is the half-power bandwidth relative to the uniform distribution).

In Figure 9.8, the patterns from two offset feeds plus one feed at the focus are shown. Also shown is the calculated sum of these three separate patterns.

From Figure 9.8, it can be seen that the overall beamwidth is a direct function of the beamwidths of the component patterns. The amplitude taper in the pillbox widened the individual beams about 16 per cent (which is why the point marked X in Figure 9.1 is shifted to the right). This means that the individual beamwidths could be reduced about 16 per cent by making the feed patterns more closely approach a uniform amplitude distribution. The gain would then be improved with little change in effectiveness of the sidelobe cancellations. Figure 9.8 also points up one of the difficulties of this approach, mainly requiring that the maxima and zeros exactly coincide. The asymmetry of the sum pattern can be attributed to the placement and to the patterns of the multiple feeds.

One of the advantages of the three feed method is that it does allow a large variety of pattern shapes to be synthesized. A very narrow beamwidth can be obtained by making the center pattern out of phase with the offset patterns. Of course, this raises the sidelobes since they now add. Or a three-beam pattern could be made by setting all three beam amplitudes equal but out of phase. A very broad beamwidth would result if they were all in phase and of equal amplitude.

Even though the three-feed method has greater potential for lowering the sidelobe levels, its complexity negates some of this gain. As Figure 9.7 shows, there was only an additional 2.5 dB gained from going to this approach (theoretically it should have been a 21 dB gain). Actually, the two-feed method gave slightly better results than anticipated, further closing the gap between the two approaches. The reason is that there was cubic phase error introduced by the two offset feeds of the parabolic reflector. Cubic phase error has the effect of raising the level of the sidelobes on the side of the main beam nearer the normal to the antenna aperture. One side is now nearer the normal because the beam has also been tilted as in the case of a linear phase error. From Figure 9.2B, the N+1 numbered sidelobe on the side toward the normal of one component pattern is subtracted from the N numbered sidelobe on the side away from normal of the other component pattern. So the cancellation obtained by subtracting an increased N+1 sidelobe from an unchanged N sidelobe will be improved. Although cubic phase error is a difficult quantity to control, it has the additional advantage of not increasing the beamwidth. Without the help of this error, the sidelobes could be expected to drop down to 22 dB below the main beam when two feeds are used as compared to 13 dB down for a single feed. The beamwidth would be increased about 30 per cent over the same antenna with a single feed.

Conclusions

The desired overall distribution sought was a cosine-squared distribution with a uniform distribution of 7 per cent of the maximum of the cosine-squared distribution added to it. The major lobes of the two distributions will add and the minor lobes will tend to cancel each other. This should bring the sidelobes down to about 43 dB below the main beam.

This method can be applied to an unmodified line source antenna that displays a fairly symmetrical, tapered amplitude distribution and approximately linear phase distribution across the aperture. The antenna must be a parabolic reflector type, and the feeds should be unobtrusive, so far as the radiation pattern is concerned. The problem of feed blocking becomes more acute as the feed system becomes more elaborate. Also, mutual coupling may become a problem when a multiple feed system is used.

The method could be extended to two orthogonal planes by using a multiple-feed system for each plane. However, the problem of aperture blocking by the feed system would increase. Sletten and Maviordes [3] have shown that even if sidelobe cancellation is achieved in the two principal planes, very little improvement will occur in the diagonal planes.

The overall pattern can be determined quite accurately by adding together the patterns obtained from the individual feeds operating alone. In this manner the contribution from each feed may be found and the optimum placement for each feed determined.

This technique may be applied to existing line-source antennas where the sidelobe level needs to be below 30 dB. The design concept is relatively simple, the main disadvantage being how close the individual feed patterns and placements can be maintained with respect to the desired values.

REFERENCES

[1] Kiely, D.G., A.E. Collins, and G.W. Evans, "Cheese Aerials For Marine Navigational Radar," *Proc. IEE* (London) part III, Vol. 98, No. 51, Jan. 1951. An investigation of the pillbox, or "cheese," antenna aimed at minimizing side lobes.

[2] Bogner, R.D., *Final Report on Investigation of Side-Lobe Reduction.* Airborne Instruments Laboratory, AD 16100.

[3] Sletten, C.J., and W.G. Maviordes, "A Method of Side Lobe Reduction," *AFCRC,* March 1952.

[4] Jasik, H., "Aperture Distributions for Low Side-Lobe Antennas," presented at the IRE-URSI Symposium at San Diego, Calif., April 3, 1950.

[5] Andrews, C.L., "Diffraction Pattern of a Circular Aperture at Short Distances," *Physical Review,* Vol. 71, No. 11, June 1947.

[6] Andrews, C.L., "Diffraction Patterns in a Circular Aperture Measured in the Microwave Region," *Journal of Applied Physics,* Vol. 21, No. 8, August 1950.

Figures 9.1 and 9.2 Kiely, D.G., A.E. Collins, and G.W. Evans, "Cheese Aerials for Marine Navigational Radar," *Proc. IEE* (London), part III, Vol. 98, No. 51, Jan. 195.

Figures 9.3 to 9.6 Jasik, H., "Aperture Distributions for Low Side-Lobe Antennas," IRE-URSI Symposium, San Diego, California, April 3, 1950.

Figures 9.7 and 9.8 Bogner, R.D., Final Report on Investigation of Side-Lobe Reduction, Airborne Instruments Laboratory, AD 16100.

10

DIELECTRIC ROD WAVEGUIDE ARRAY

A dielectric rod waveguide array is composed of a slotted waveguide array with dielectric rods inserted in the slots to reduce the sidelobe levels. Each dielectric rod can be thought of as a polyrod antenna. The guiding effect of a high dielectric constant rod is a result of the property that electromagnetic energy tends to stay in the optically denser medium. At the boundary between the rod and the surrounding medium, most of the incident energy on this interface is reflected back into the rod.

The pattern of an array of such radiators would then be the array factor multiplied by the individual element pattern.

The dielectric rod acts as a circular waveguide operating as an endfire radiator. The lowest mode of propagation is the hybrid TE_{11} mode. To support this mode, the diameter D of the waveguide must be at least $0.58\sqrt{\xi r}$, where λ_o is the free-space wavelength and ξr is the relative dielectric constant. However, since the spacing of the elements in the array must be less than a free-space wavelength, it is necessary to make the diameters as small as possible. Increasing the relative dielectric constant reduces the required diameter. Figure 10.1 shows the ratio of free-space wavelength to dielectric rod wavelength as a function of rod diameter for various values of dielectric constant. The higher the ratio of free-space wavelength to dielectric rod wavelength the greater will be the guiding effect of the rod. However, reflection at the end of the rod will also increase with a greater ratio, giving rise to a greater VSWR.

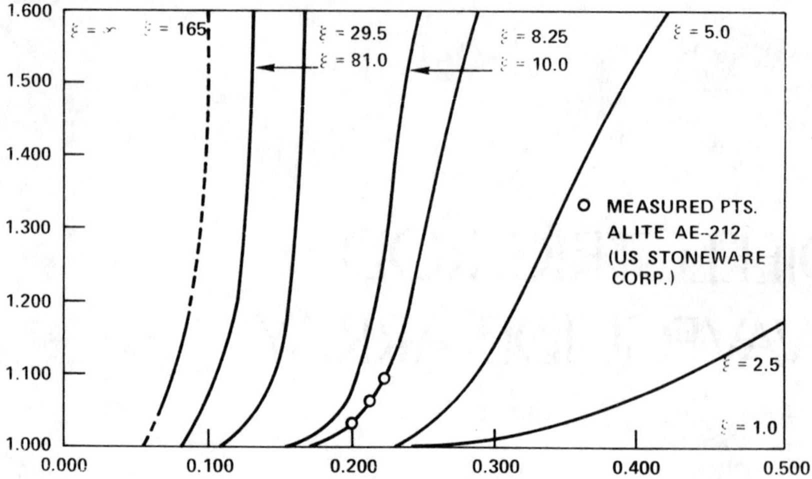

Figure 10.1 Ratio of free-space wavelength to the dielectric rod wavelength as a function of rod diameter

An approximate design base was established from Kielys [1] analysis and patterns were taken [2] on various configurations of R-1 ferrite rods which generally agreed with the predicted results. The ferrite rods [3] had a relative dielectric constant of 13.6 and a permeability of 0.76. In Figure 10.2 the radiation pattern of a ferrite rod 0.27 inch in diameter and 5.4 inches in length, operated at a frequency of 9500 MHz is shown. For comparison, the pattern of an unmodified slot is also shown. The ferrod exhibits a much better off-axis sidelobe rejection capability than the slot.

One method of coupling the polyrod to the waveguide slot is through an auxiliary waveguide. This is done to provide a better impedance match between the polyrod and the waveguide than could be obtained by directly inserting the polyrod into the slot in the waveguide. Since data are available on matching one waveguide to another, the only problem left is matching the polyrod to the auxiliary guide. The method used in one study [4] was to use irises of varying thicknesses and different dielectric constants to provide to the incident wave a gradually tapered dielectric constant. This technique is shown in Figure 10.3.

An alite rod radiator was operated in the hybrid TE_{11} mode when excited from a 1 inch by 0.5 inch rectangular waveguide operated in the fundamental TE_{01} mode. The E-plane beamwidth is

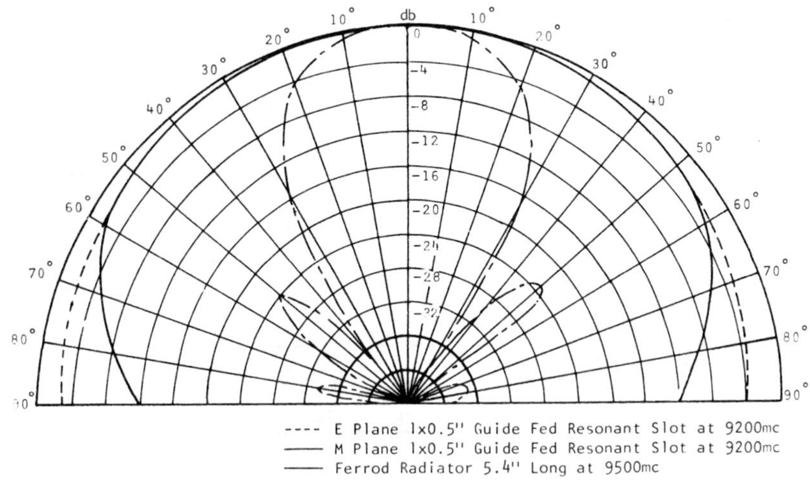

Figure 10.2 Comparison of experimental ferrod and slot radiation patterns at X-band.

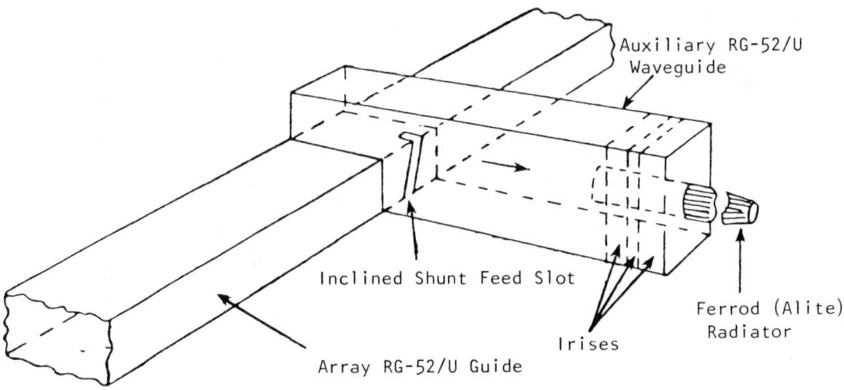

Figure 10.3 Auxiliary waveguide method for coupling polyrod to waveguide slot

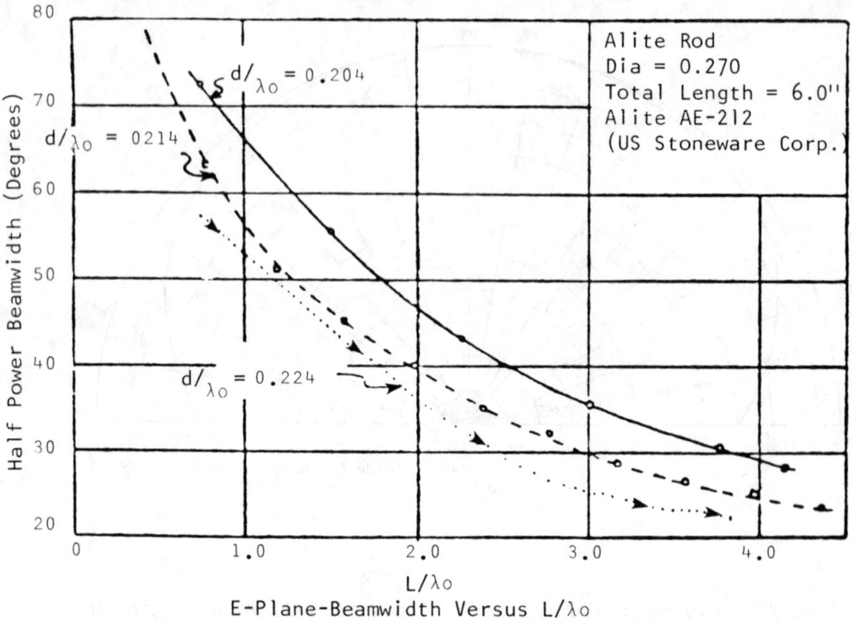

Figure 10.4 E-plane beamwidth versus L/λ_o

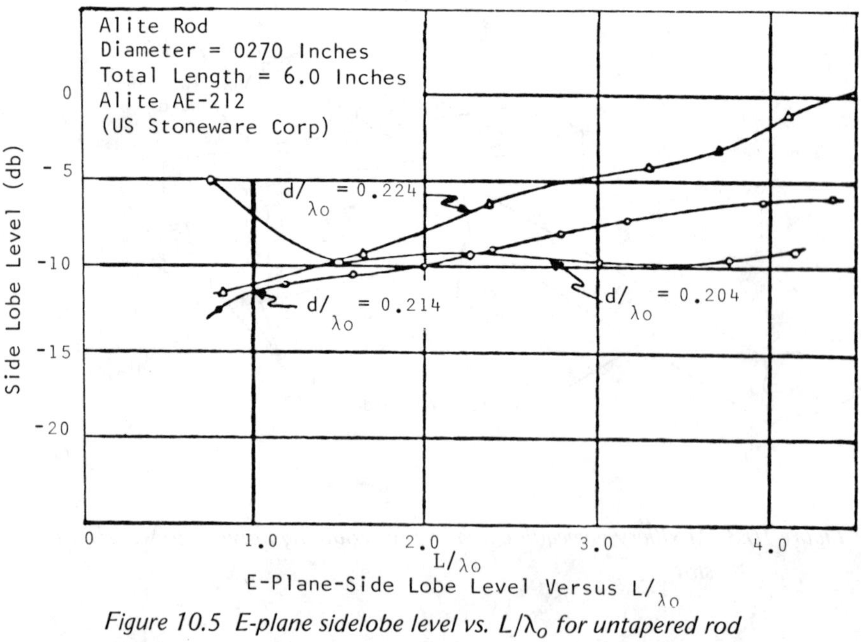

Figure 10.5 E-plane sidelobe level vs. L/λ_o for untapered rod

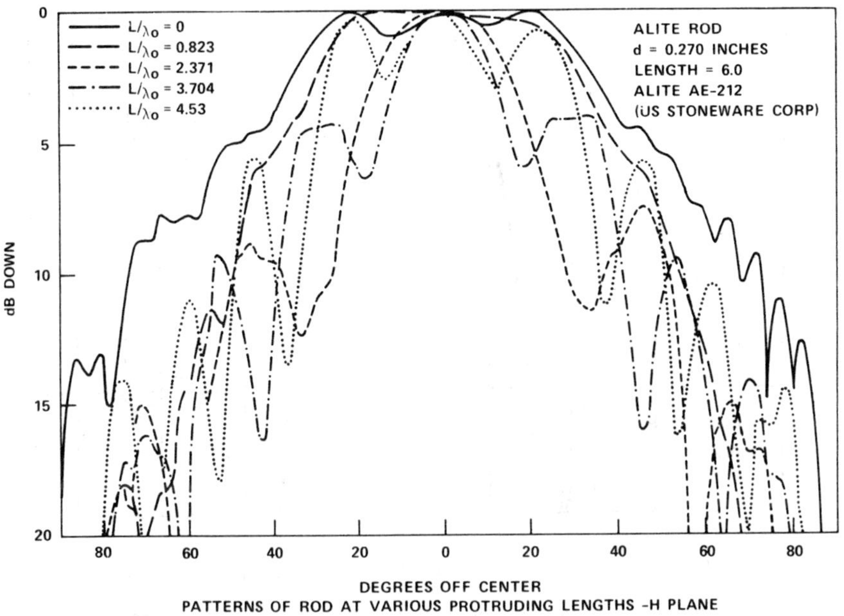

Figure 10.6 H-plane patterns for various rod lengths

plotted versus the polyrod length in wavelengths in Figure 10.4. The rod was untapered and the beamwidth is decreased as the length is increased. The beamwidth is also improved if the diameter of the rod is increased. This can be thought of as two linear arrays being moved further apart. The opposite sides of the rod correspond to the two arrays. This modifies the general (sin X)/X pattern of the rod by the factor: $\cos(\pi d/\lambda)\sin\theta$ where d is the diameter of the rod, and θ is the off-axis angle.

Figure 10.5 shows the sidelobe levels of the rods as a function of L/λ_o. The curve of sidelobe level versus L/λ_o for $d/\lambda_o = 0.224$ shows that the sidelobe is at about the same level as the main beam when L/λ_o is 4.0 or greater. The Hansen-Woodyard [5] criterion is being exceeded for these lengths and any further increase in length would widen the beam split. The Hansen-Woodyard criterion is a method of phasing an end-fire antenna for increasing its directivity.

Figure 10.6 indicates the radiation patterns obtained for various values of L/λ_o for untapered rods of D/λ_o of 0.224. The high sidelobes appear for rod lengths exceeding $4\lambda_o$.

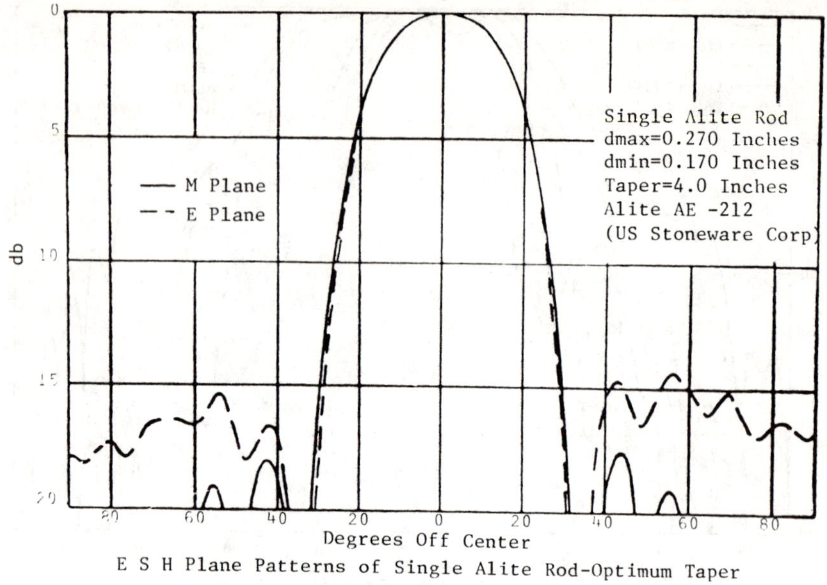

Figure 10.7 E-plane beamwidth as a function of percent taper

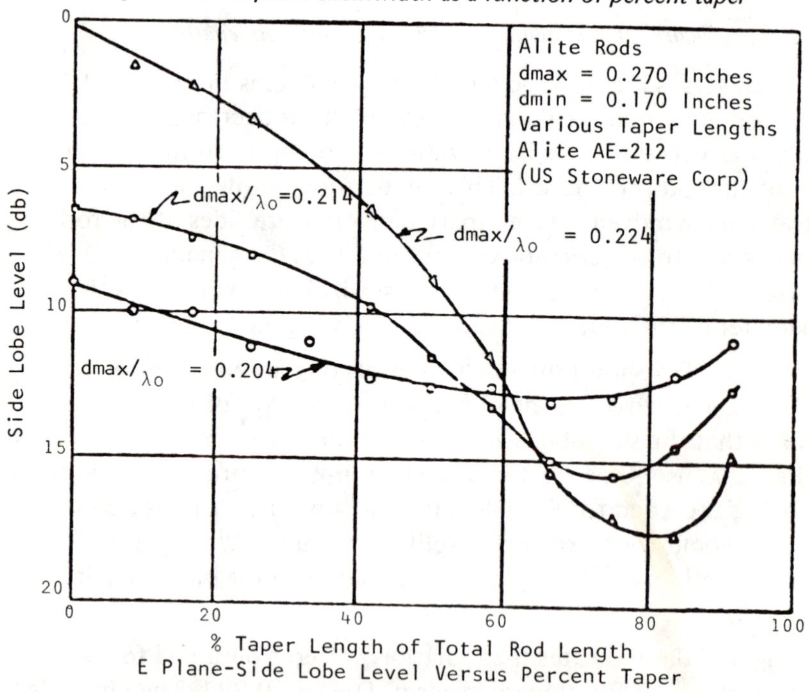

Figure 10.8 Sidelobe level versus percent taper

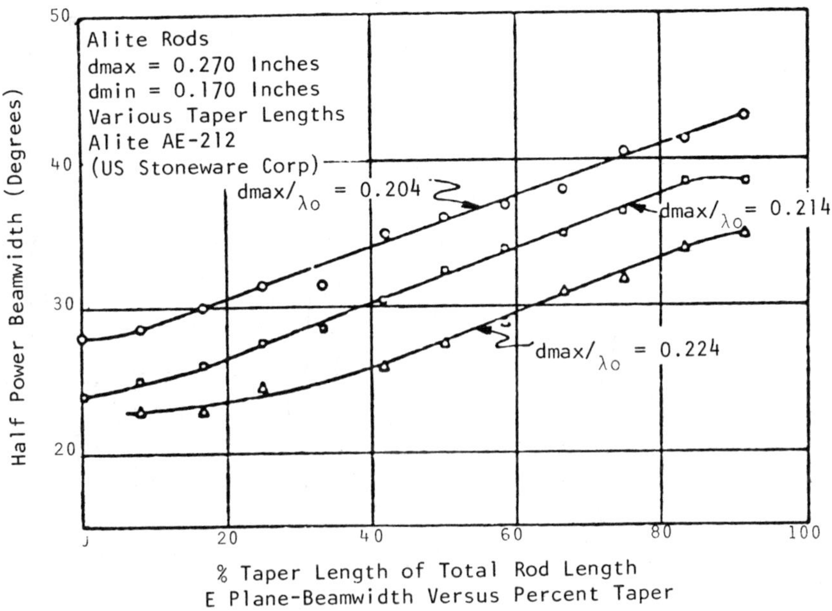

Figure 10.9 E and H plane patterns of optimally tapered rod

The high sidelobe levels of the untapered rods can be lowered by tapering the rods. There is no point in reducing the diameter beyond the point where the ratio of free-space wavelength to dielectric rod wavelength is about equal to one. Independent work by Kiely [1] and Mallach [6] indicate the best ratio of minimum to maximum rod diameter is 0.63. Figure 10.7 shows the effect on half-power beamwidth of varying the per cent of taper on a $4.5\lambda_o$ alite rod.

Figure 10.8 indicates how the sidelobe levels change as a function of per cent taper. The best sidelobe level obtained was −17.5 dB at 80 per cent taper from the $0.224\lambda_o$ rod. The halfpower beamwidth for this case was 34 degrees.

Figure 10.9 is a typical radiation pattern of a tapered alite rod. This pattern was taken at 9375 MHz and shows the nearly similar E and H plane beams resulting from the similar excitations of the hybrid mode. A VSWR of less than 1.5 was easily achieved over a ± 5 percent bandwidth.

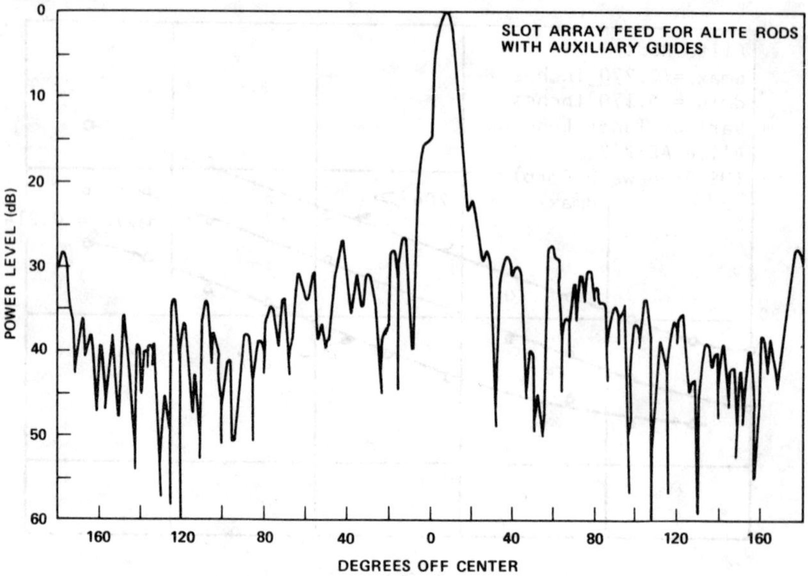

Figure 10.10 E-plane pattern of 17 element array with auxiliary guides

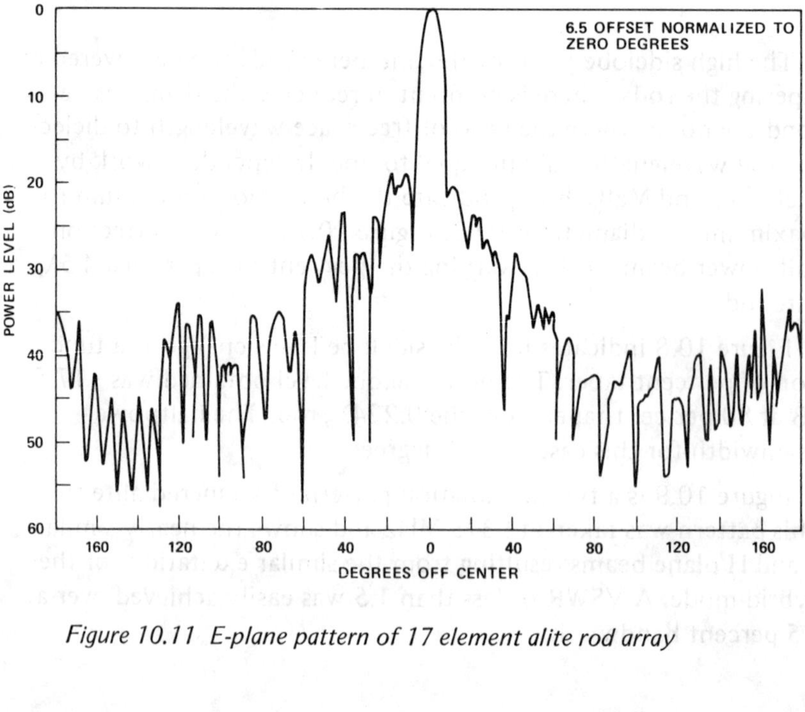

Figure 10.11 E-plane pattern of 17 element alite rod array

A seventeen element alite rod array was built and tested at 9375 MHz. The spacing between elements was 0.8 inches and each rod had the following dimensions:

Maximum diameter = 0.27 inches
Minimum diameter = 0.17 inches
Length = 5.5 inches
Length of Taper = 4.0 inches

A Taylor aperture distribution was used to provide an array half-power beamwidth of 6.35 degrees. This distribution was provided by a standard RG-52/U waveguide with inclined shunt slots. Auxiliary guides coupled the energy from the slots to the polyrod radiators.

Figure 10.10 shows the E-plane pattern of the seventeen element array without the polyrods inserted in the auxiliary waveguides. The far-out sidelobes are fairly high, being above -35 dB much of the time. The backlobe for this array is about -26 dB.

When the polyrods are inserted into the waveguides, the pattern shown in Figure 10.11 is obtained. The half-power beamwidth is 6.5 degrees and the first sidelobes are down -18 dB and -20 dB respectively, as compared to -15.5 dB and -25 dB for the pattern in Figure 10.10. The far-our sidelobes have been reduced marginally over those of Figure 10.10 but the backlobe has been lowered about 7 dB.

Conclusion

For the particular array tested, scant improvement was obtained by employing the dielectric rods. However, it was felt that the first sidelobes could have been maintained below -25 dB and the far-out sidelobes below -40 dB by the use of proper aperture control.

SUMMARY OF ANTENNA TECHNIQUES

The techniques described herein generally fall into one of three areas. Those areas are:

(1) Modifications that can be done to the site, irrespective of the antenna employed on the site;
(2) Modifications to parabolic reflector antennas, since they are of primary concern; and
(3) Antenna configurations, other than parabolic types, that are capable (or at least have the potential) of exhibiting low sidelobes and backlobes.

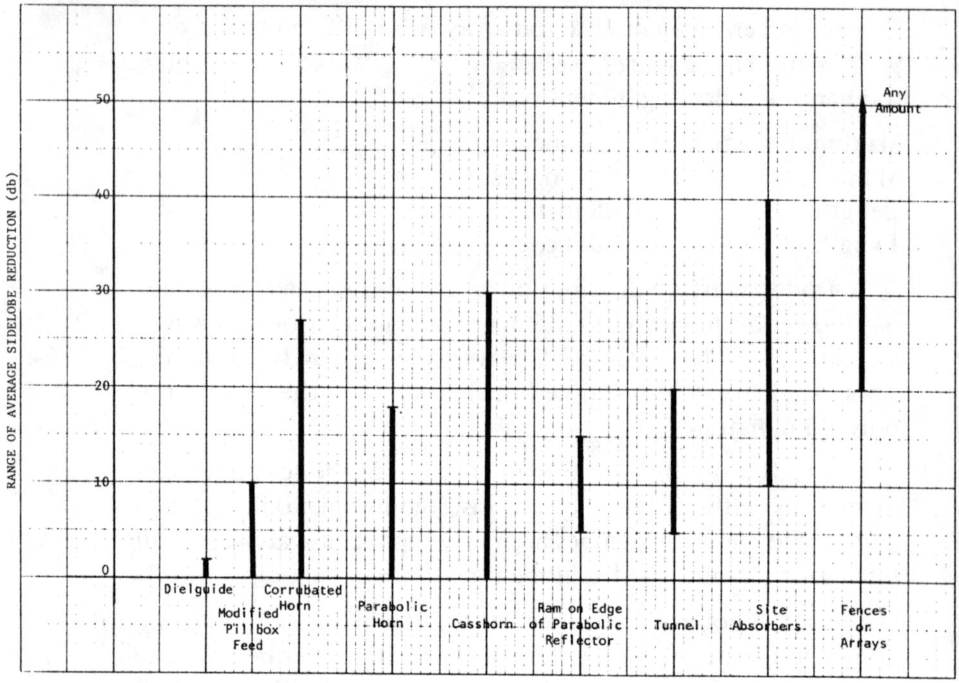

Figure 10.12 Anticipated sidelobe reduction due to various techniques

The general ranges of improvement that may be attained by the use of the techniques is illustrated in Figure 10.12. It should be borne in mind that, in many cases, the techniques can be combined. For example, the dielguide feed can be illuminated by a corrugated feed horn and these, in turn, can illuminate a tunnel antenna, which could be installed on a RAM modified site. It is therefore rather difficult to generalize on the total improvement that may be attained in a specific case.

The question of what may be done to typical existing co-sited highly directional antennas is one of great interest. Table 1 contains a summary of the modifications that may be used and an estimate of benefits to be attained. The specific problems should be evaluated empirically.

The improvement to be attained using techniques such as we describe should be based on the specific problems of interest. When the problem is one of sidelobe and backlobe interference between multiple sites the improvement to be achieved by the appli-

TABLE 10.1

Summary of Appropriate EMC Modifications for Cassegrain or Parabolic High Gain Antennas

	Modification	*Remarks*
Site Modifiers	1. Application of RAM	Can reduce sidelobe median gain about 10 dB, increase standard deviation about 4 dB for high gain antennas in congested installations.
	2. Use of Fences	Intercepts energy between antennas, reduces coupling as a function of height and edge treatment.
Feed Modifiers	1. E-Plane Corruged Feed Horn	Small Horn: backlobe reduced 10 dB, 3 dB bandwidth narrowed.
		Large Horn: backlobe 57 dB down from main beam, Improvement over uncorrugated horn; 0°: 0 dB, 30°: 12 dB, 60°: 20 dB, 90°: 30 dB, 120°: 25 dB, 180°: 27 dB.
	2. Dielguide Feed (Cassegrain)	Improves efficiency and main beam gain, reduces first sidelobes slightly; almost no effect on other sidelobes or backlobes.
Reflector Modifiers	1. Tunnel Antenna	Primary effect beyond first sidelobe. Possible results: 1st sidelobe: 30 dB down, 25-90°: 50 dB, 90°-180°: 65 dB down.
	2. RAM Applied to Reflector Surface (Tapering)	5 to 10 dB reduction in sidelobes is possible accompanied by a slight reduction of effectiveness.

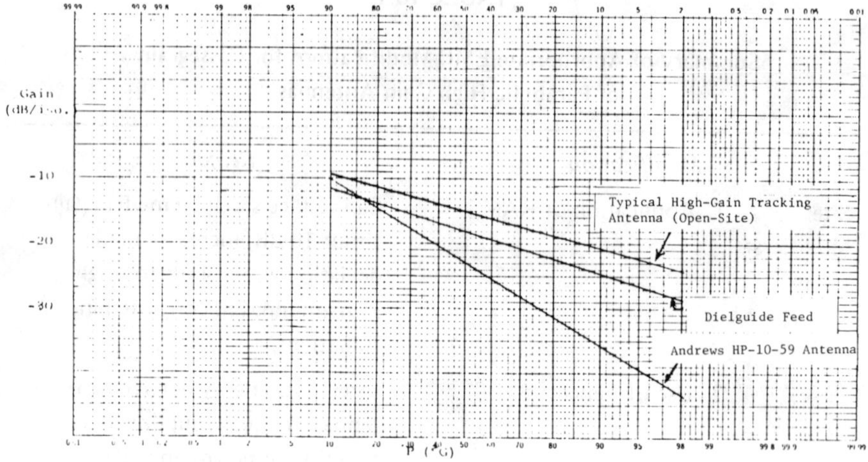

Figure 10.13 Comparison of probabilistic gains of typical high-gain antennas

cation of RAM may best be assessed by evaluating the statistics of the situation. The problem on a single site may require a more deterministic approach. For example, if a closely-positioned whip antenna is causing distortion on magnetron pulsing in a radar, the application of a sheath of frequency selective RAM to the whip may eliminate this problem, but not affect the inter-site compatibility problem to any significant degree. When the main problem is interference between two tracking radars on a single site, the use of a fence may be an effective way of decoupling the two radars.

Figure 10.13 shows the typical reduction of sidelobe and backlobe gain as a result of two modifications.

REFERENCES

[1] Kiely, D.G., *Dielectric Aerials,* Methuen's Monographs Series, Bath, England, March 1952.

[2] Reggia, F., E.G. Spencer, R.D. Hatcher, and J.E. Tompkins, "Ferrod Radiator Systems," Tech. Rep. No. TR-357, Diamond Ordnance Fuze Lab., June 1956.

[3] General Ceramics Corporation R-1 Ferrite.

[4] "Final Report on Theoretical and Experimental Investigations on Methods of Reducing Antenna Sidelobes," Melpar, Inc., AD 152-394.

[5] Hansen, W.W., and J.R. Woodyard, "A New Principle in Directional Antenna Design," *Proc. IRE,* March 26, 1938.

[6] Mallach, P., "Dielectric Directional Antennas for DM and CM Waves," Air Material Command Report, F-TS-2223-RE, February 1948.

[7] Kurtz, L.A., and E. Strumwasser, "Design Application of Edge Slots," Tech. Memo No. 226, R&D Lab., Hughes Aircraft Company, January 1952.

[8] Lee, S.W., and R. Mittra, "Radiation from Dielectric Loaded Arrays of Parallel-Plate Waveguides," *IEEE Trans.,* Vol. AP-16, September 1968.

[9] Galindo, V., and C.P. Wu, "Dielectric-loaded and Covered Rectangular Waveguide Phased Array," *Bell Systems Tech. J.,* Vol. 47, January 1968.

[10] Lewis, B.L., "Large Diameter Dielectric Rod End-Fire Antennas," *IEEE Trans.,* Vol. AP-14, March 1966.

[11] Snyder, A.W., R.L. Kyhl, "Surface Mode Propagation Along An Array of Dielectric Rods with All Elements Excited Identically," *IEEE Trans.,* Vol. AP-14, July 1966.

[12] Peters, Ruccuck, and Du, "RFI Reduction by Control of Sidelobes," *IEEE Transactions on Electromagnetic Compatibility,* Vol. EMC-6, January 1964, pp. 1-11.

[13] Johnson, R.C., "Mutual Gain of Radar Search Antennas," Ninth Tri-Service Conference on Electromagnetic Compatibility, October 1963, pp. 368-373.

[14] Caldecott and Brown, "The Design of Low Noise Antennas for Tropospheric Scatter Reception," Antenna Lab., Ohio State University Research Foundation, January 15, 1960, Report 904-1.

[15] "Final Report on Theoretical and Experimental Investigations on Methods of Reducing Antenna Side Lobes," Melpar, Inc., June 15, 1958.

[16] "Interference Notebook," Rome Air Development Center, RADC-TR-66-1.

Figures 10.1 and 10.2 Reggia, F., et al., Ferrod Radiator Systems, Tech. Rep. No. TR-357, Diamond Ordnance Fuze Lab., June 1956.

Figures 10.3 to 10.10 Final Report on Theoretical and Experimental Investigations on Methods of Reducing Antenna Side-Lobes, Melpar, Inc., June 15, 1958.

11

METHODS of MEETING EMISSION LIMITS for TRANSMITTERS

The overall emission spectrum can be broken down into three major parts:

a. Out-of-band spurious and harmonic radiation
b. In-band modulation
c. In-band spurious emissions.

The major controlling factors determining the levels of the emission spectrum at the antenna terminals for these three regions are the pulse shape, the power output tube, and the frequency response of the RF section between the transmitter and the antenna.

PULSE SHAPING

Figure 11.1 shows the relative decrease in the power spectrum envelope as a function of frequency separation for certain pulse shapes of interest. The relationships of the central spectral power density and the peak power are contained in Table 11.1. It is highly significant to note that for the rectangular waveshape, assuming typical search radar parameters, i.e., 4 μsec pulse width and a PRF of 400 Hz, the maximum radiated peak power that could be used without exceeding a given emission bandwidth requirement for a radar transmitter with an antenna whose main beam gain is 30dB, is about 50 dBm (100 watts). If a cosine squared pulse is used under the same conditions, 130 dBm peak power could be used, provided that the mathematical promise of pulse shaping could be matched by the physical methods available. It is doubtful that any future radars, including satellite and missile surveillance and tracking radars, will exceed 100 MW (110 dBm); therefore, by applying

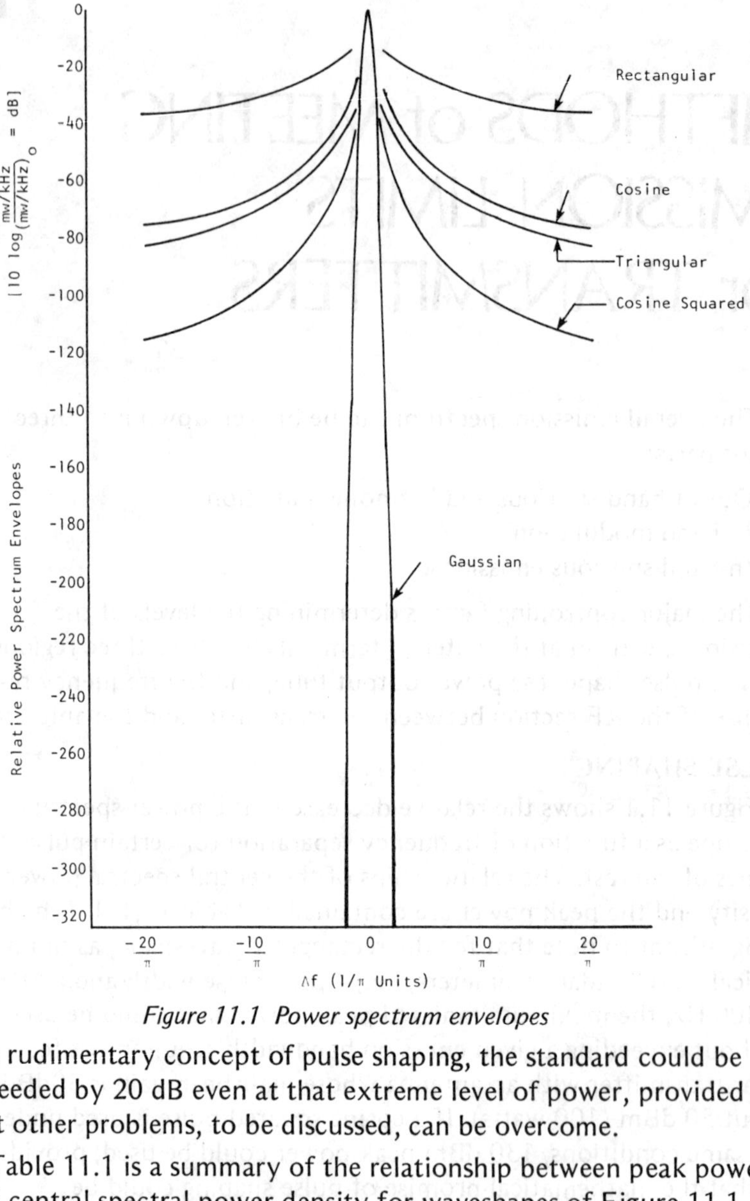

Figure 11.1 Power spectrum envelopes

this rudimentary concept of pulse shaping, the standard could be exceeded by 20 dB even at that extreme level of power, provided that other problems, to be discussed, can be overcome.

Table 11.1 is a summary of the relationship between peak power and central spectral power density for waveshapes of Figure 11.1. Note that, for practical purposes, the relationship, $P_s(f_o) = (P\tau^2)/T\ (10)^3$, is an adequate approximation for all pulse shapes, where P is the peak power (mw), $P_s(f_o)$ is the spectral power density at the spectrum center (mw/kHz), T is the period (sec), and τ is the pulse duration (sec) (τ = time between half power points).

TABLE 11.1

Central Spectral Power Density as a Function of Peak Power

Rectangular Pulse:

$$P_s\,(f_o) = \frac{P\tau^2}{T}\,(10)^3$$

Triangular Pulse:

$$P_s\,(f_o) = 2.914\,\frac{P\tau^2}{T}\,(10)^3$$

Cosine Pulse:

$$P_s\,(f_o) = 1.621\,\frac{P\tau^2}{T}\,(10)^3$$

Cosine² Pulse:

$$P_s\,(f_o) = 1.886\,\frac{P\tau^2}{T}\,(10)^3$$

Gaussian Pulse (Non-Realizable):

$$P_s\,(f_o) = 2.266\,\frac{P\tau^2}{T}\,(10)^3$$

where: P = peak power (mv)

$P_s(f_o)$ = spectral power density at spectrum center (mw/kHz)

T = period (sec.)

τ = pulse width (sec.), time between half power points

While considered the optimum, the Gaussian waveform is theoretically unachievable due to its infinite skirts. Much work has been done on an approximation to th's waveform known as the "Binominal" pulse [1]. While it may not presently be necessary to shape pulses to that extent, it is apparently required that some degree of pulse shaping be employed in order to conserve the spectrum.

HIGH POWER OUTPUT STAGES

High power pulse radars primarily employ either magnetron, amplitron, klystron, TWT, Triode, or Tetrode power output tubes. Since the magnetron is an oscillator, it is not amenable to pulse-shaping techniques to any great extent. In addition it usually creates a phase shift in the RF pulse due to "pushing", causing an asymmetrical spectrum. The external cavity magnetron, such as the Stabilitron, does provide a more symmetrical spectrum, but this, in itself, is not sufficient to meet the standard at high pulse power levels encountered in most high power radars. The magnetron is a light weight and economical approach to high-power microwave generation, but its limitations in terms of spectrum control by pulse shaping and excessive generation of spurious emissions make it a very poor choice for EMC.

Non-harmonic spurious outputs from magnetrons appear to be a function of the coupled resonant circuits inherent in the magnetron structure. For N coupled resonant circuits or cavities, there are N fundamental modes of resonance which are, in general, different from each other and from the resonant frequency of the individual uncoupled resonators. In actuality, the zero mode cannot be readily excited and, for a symmetrical resonant system, there will only be N/2 principal modes. The other (N/2) - 1 modes are degenerate modes. The π mode is the one generally used for design purposes. There are, in addition, secondary modes of operation as well as the principal spurious mode. These are sometimes due to mixing of the principal modes. Computing the frequencies of spurious outputs due to these causes is possible, but is usually not a very fruitful exercise since no method has been established for determining the coupling factor for a particular mixing product. Only a small fraction of the possible products are coupled efficiently to the output.

One attempt at modelling magnetron spurious emissions on a statistical basis, as a function of frequency difference from the fundamental, concluded that the frequency statistic for magnetron non-harmonic outputs can be represented by a uniform distribution [2]. A method that might suppress such spurious signals is the use of ferrite mode suppressors in the magnetron.

The other devices mentioned are all amplifiers and therefore could be used with pulse-shaping circuitry. Certain of these are

TABLE 11.2

SUMMARY OF RADIATED RADAR HARMONIC LEVELS
(dB BELOW FUNDAMENTAL)

Harmonic Number / Tube Type	2	3	4	5	6	7	8	9	10
MAGNETRON									
Mean	78.1	71.7	77.1	86.0	87.2	91.9	99.3	–	–
No. of Samples	77	59	34	23	17	7	–	–	–
KLYSTRON									
Mean	71.3	78.2	76.9	73.9	82.3	87.2	–	–	–
No. of Samples	44	27	21	8	7	4	–	–	–
TETRODE & TRIODE									
Mean	83.2	76.0	99.6	90.0	96.7	100.2	106.2	97.2	100.4
No. of Samples	14	13	10	11	9	9	3	4	2

*Systems with Waveguide Transmission Lines Only.

susceptible to non-harmonic spurious emission problems as the magnetron (e.g., the amplitron), and for the same reasons.

The harmonic output of all these devices have been found to be similar, as shown in Table 11.2. This table is a compilation of measured spectrum signature results from 34 radar types [3]. It should be noted that all harmonics from a radar are not necessarily generated in non-linearities in the tube, but may be generated in other non-linear elements such as the duplexer. It is also important to consider that the radiated power of the harmonics generated in the power output tube is usually considerably reduced by the RF circuitry between the tube and the antenna.

When space and weight permit, the klystron and TWT offer many advantages from an interference viewpoint. Measurements in both the field and laboratory have shown that klystrons rarely produce spurious signals of appreciable strength in the near-band. The standard klystron does produce significant harmonic output equal to that of the magnetron (20 to 50 dB below the fundamental at the tube output and 50 to 80 dB below at the antenna terminals). A new klystron under development (the axially symmetric cavity klystron) is said to offer 40 dB better performance at harmonic levels up to the fourth harmonic [4]. It has also been noted that a 5 to 10 dB reduction in harmonic output can frequently be obtained by filtering the low level drive to the tube [5].

Travelling wave tubes show a tendency to generate in-band spurious signals due to "band edge" oscillations. The harmonic output is somewhat higher than klystrons but in the same general range (20 to 40 dB below the fundamental at the tube output).

The amplitron is a crossed field amplifier and can be thought of as a hybrid of a magnetron and a TWT. As a result it shows some of the undesirable behavior of both in regard to near-band spurious and band edge oscillations.

Triode oscillators are sometimes used in VHF and UHF radars. These tubes are fairly clean, although susceptible to the problem of "dark noise", i.e., the emission of noise resulting from residual plate current when the tube is cut off. With plate voltages on the order of 25 KV, even a current of a few milliamperes can cause severe desensitization of nearby receivers [6].

An additional source of undesirable radiation in radar transmitters is the modulator pulse used to drive a magnetron; it is rich in harmonics due to its sharp rise and fall time, and can radiate directly unless precautions are taken in shielding the modulator. This is another reason to avoid magnetrons when trying to achieve EMC.

In summary, the application of the techniques of pulse shaping should enable the near-in spectrum to meet the limits of the standard provided that non-harmonic emissions can be suppressed. In order to adequately shape pulses, the use of amplifiers rather than oscillator-amplifier (magnetron) output devices is required.

In regard to in-band spurious emissions, whose generation mechanism is presently not fully understood, the magnetron and amplitron are the biggest offenders. The TWT and amplitron are subject to band-edge oscillations. In terms of harmonic emissions, the levels of presently available tubes appear equivalent. The ranking of these devices in regard to EMC is shown in Table 11.3.

TABLE 11.3

Order of Preference of Output Tubes in Regard to Interference

Tube Type	*Remarks*
1. Axially-symmetric cavity klystrons	Not fully developed or available
2. Klystrons	Relatively spurious-free
3. Triodes	Susceptible to "Dark Noise"
4. TWT's	Band-Edge Oscillations
5. Amplitrons	Band-Edge Oscillations and Spurious Problems
6. Magnetrons	Limited Pulse Shaping and Spurious Problems

SOLID STATE DEVICES

The development of solid state devices suitable for high power radars is still at least 4 to 5 years in the future, although low power radars using these devices are now under development. Data available at this point indicate that solid state devices such as GaAs oscillators are inherently noisier than "hollow-state" devices (tubes) with spurious and harmonic levels approximately 30 dB below the fundamental power. Part of this problem is due to noise caused by electron collision with the lattice structure. Studies to determine methods of cleaning up the spectrum of these devices without filtering have been largely unsuccessful.

In order to generate large amounts of power, such fabrication techniques as stripline construction will have to be employed, and the systems will probably be phased arrays. The next subject to be discussed will be methods for suppressing out-of-band spurious and harmonic emissions.

FILTERS

The suppression of radar transmitter out-of-band energy generally leads to consideration of the use of filters. The characteristics and parameters that must be considered in the application of filters for this purpose are: (1) absorptive vs. reflective type, (2) band-pass, low-pass, high-pass or band stop type, (3) pass-band VSWR, (4) stop-band VSWR, (5) insertion loss, (6) waveform distortion, (7) stop-band loss, (8) multimode response, and (9) size and weight.

Adding any additional loss, cost, or weight to a system must be carefully evaluated. Therefore, it might be stated at the outset that radar designers and EMC engineers both prefer not to use filters if their use can be avoided. However, filters have proven to be the only dependable, effective way to suppress out-of-band signals up to the present time. Young [7] provides the following three cautions to systems designers:

1. Actual filter performance in a system may be different than the filter operating by itself, due to mismatches between the generator and the load.
2. An absorbing filter is preferable to a reflecting filter. Thus a leaky waveguide filter (which absorbs in the stop-band) is preferable to a waffle-iron filter (which reflects in the stop-band) if both have the same electrical performance between matched terminations. The waffle-iron filter is much more compact, but a reflecting filter in a high-power system may cause arcing due to resonance effects and could damage the transmitter. (This can be avoided by using directional coupler or leaky wall filter pads, or by placing the waffle iron filter in the tube itself for systems operating at X-band and above.)
3. The filter must be considered as part of a whole system and applied accordingly. For example, if a gas-discharge duplexer is used in a radar, and if this duplexer generates appreciable harmonic power, it will not help much to place a good harmonic filter between the transmitter and the duplexer. (One of the most prevalent land-based radar harmonic problems was caused by this phenomenon.

The power handling capability of a filter is determined by the conditions for dielectric breakdown caused by high electric fields. One example of the relationship of power handling capacity as a function of the dielectric is illustrated in Table 11.4.

The pass-band VSWR and insertion loss are important because high values lead to heating and power breakdown, and reduce the overall system capability. Values considered acceptable and obtainable are an insertion loss of 0.2 dB or less and a VSWR of 1.2:1 or less over a bandwidth of 10 to 15 percent.

Waveform distortion is an important consideration for in-band band-pass filters. Since this approach to in-band spectrum control

TABLE 11.4

Power Handling Capacity as a Function of the Dielectric*

Power Handling Capacity	*Gas*	*Pressure*
400 KW	Air	0 psig
1 MW	Air	15 psig
6.5 MW	SF6	15 psig
12 MW	Freon 114	15 psig

*After McCable and Stone, *Survey of High-Power Microwave Filters,* IEEE Transactions on Aerospace and Electronic Systems, October 1965.

severely limits frequency flexibility, it will not be considered here. The effect of waveform distortion on out-of-band emissions is insignificant.

To be effective at waveguide frequencies where many modes of propagation may be supported, a filter must provide high stop-band attenuation to all propagation modes. The response of a filter to spurious energy may be to reflect it, absorb it, convert it to another mode, or transmit it in any mode. The insertion loss in the stop-band of a high power filter is the amount of attenuation of the incident energy to the energy leaving the filter, even though the modal distribution may be different.

The stop-band VSWR becomes important when the reflected energy is of sufficient magnitude to cause arcing across the power output tube window or to create a resonant condition which, when added vectorially to the fundamental signal, will cause RF breakdown. As a rule-of-thumb, it is reasonable to require that the stop-band VSWR of absorptive filters be equivalent to the harmonic number, e.g., 2.0:1 at the second harmonic, 3.0:1 at the third harmonic, etc. Reflective, or reactive, filters reflect all the suppressed energy back towards the generator.

Band pass filters are most commonly of the cavity resonator type. A tunable three cavity L-band (1250-1350 MHz) filter was developed for radar use to limit emission spectrum sidebands [8]. Several fixed tuned waveguide resonant cavity filters have been developed. One of the most objectionable features of these filters, in addition to waveform distortion and limited tuning flexibility,

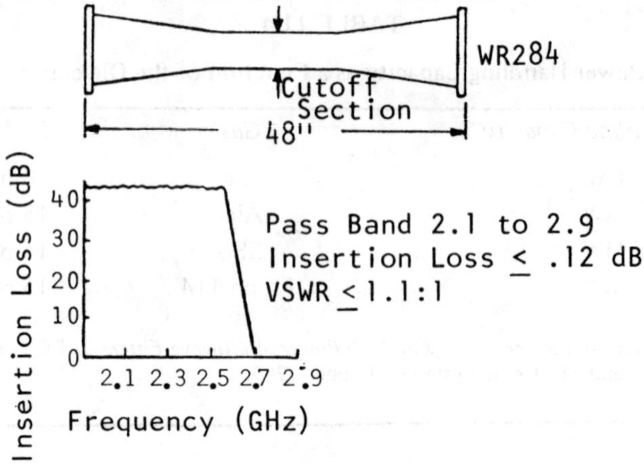

Figure 11.2 Typical S-band high-pass filter. Based on waveguide cutoff.

is that they are completely reflective, having a high stop-band VSWR. However, branching techniques can be used to absorb the energy [9]. In addition, these filter types usually have many higher frequency pass bands, even in the dominant mode range.

A number of low-pass and high-pass filters have been designed to suppress undesired signals near the fundamental signal. High-pass filters are usually not of extreme interest to the problem area under discussion because the waveguide (below cut-off) acts as a high-pass filter, and in addition, sections of waveguide can be tailored to meet high-pass filter requirements (see Figure 11.2). Such devices are, however, reflective.

Waffle-Iron Filters

The most common and useful type of high power microwave filters are the low-pass and band-stop type. One of the most significant of these is the "waffle-iron," a wide stop band, reactive filter of compact design. Typical waffle-iron filters provide 0.2 dB insertion loss in the pass-band and VSWR of about 1.2:1. Attenuation in the stop-band is usually in excess of 55 dB. The stop-band can be widened by cascading waffle-iron filters of different stop bands. Additional power handling capability can be obtained by paralleling a number of filter elements. (One example of a four-layer waffle-iron has an unpressurized power handling capability of 4MW [10]. Their prime disadvantage is that they are reactive and therefore have a high VSWR in the stop band. These filters

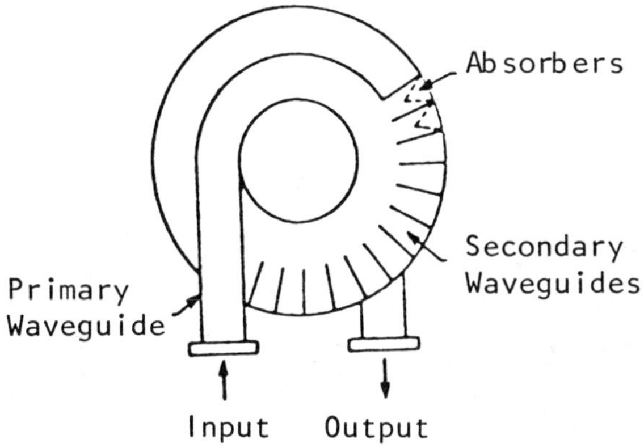

Figure 11.3 Circular dissipative-wall filter

have been integrated into high power tubes in order to suppress harmonics. This configuration eliminates the problem of reflections causing arcing, but is limited to the higher microwave frequencies due to considerations of physical size [11].

Leaky-Wall Filters

One of the earliest absorptive low-pass filters is the leaky-wall or dissipative-wall filter. This is simply a primary waveguide to which a number of secondary waveguides are connected by means of a coupling aperture. The waveguides are terminated by radar absorbing material which constitutes a wide band matched load. In the pass-band, the signals are below cutoff for the smaller secondary waveguides but at higher frequencies the side waveguides support propagation, coupling this undesired energy into the absorbing material where it is dissipated.

Typical leaky-wall filters can handle up to 6 MW without pressurization, have stop-band losses of 70 dB, pass band loss of 0.1 dB and VSWR of 1.1:1. Their most serious problem is a fall-off of stop band loss at harmonic levels above the 4th harmonic due to the "optical beaming" phenomenon; in addition, they are relatively large size devices. The optical beaming effect can be corrected by using a circular leaky-wall filter, which has a moderate stop-band loss at the lower harmonics and an increasing loss at the higher harmonics, in cascade with a straight line leaky wall filter. This makes for a heavy and bulky system. A typical circular leaky wall filter is shown in Figure 11.3.

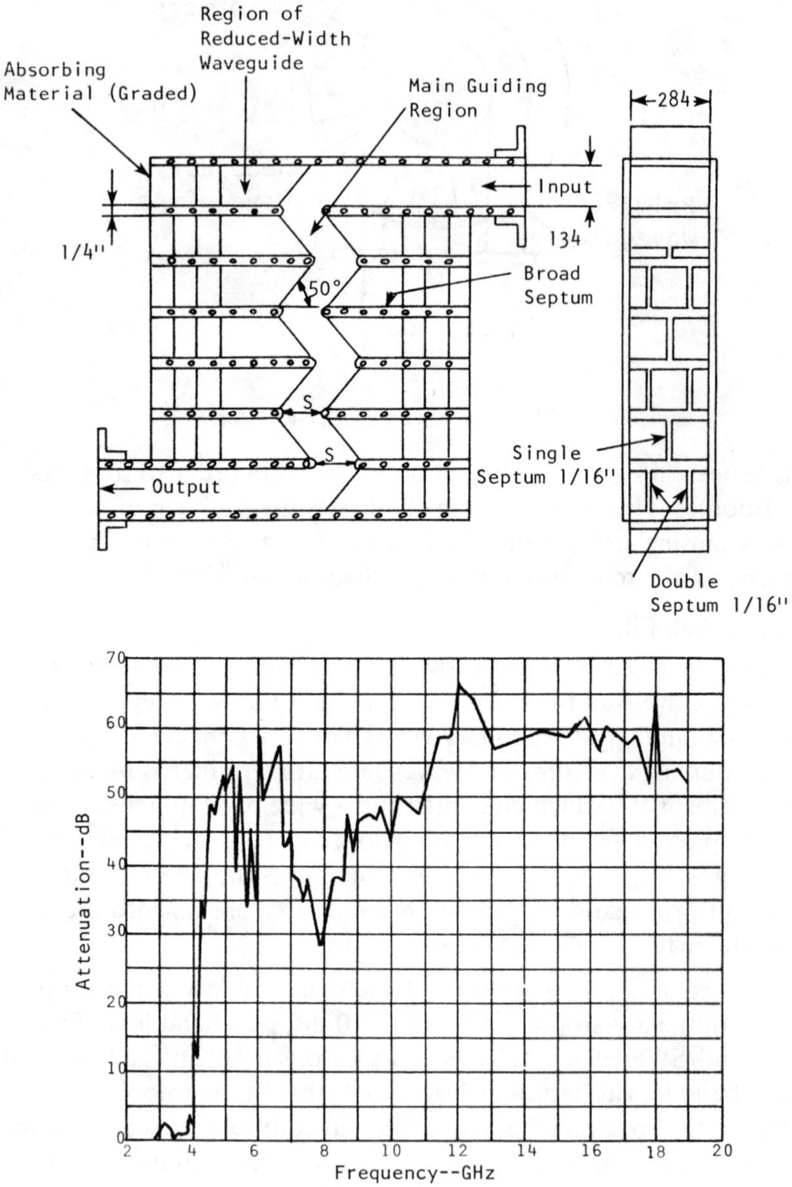

Figure 11.4 Zig-zag low pass dissipative filter

One methodof avoiding these problems is the"zig-zag"leaky wall filter shown in Figure 11.4. The optical beam effect is avoided by the zig-zag waveguide structure connecting the input and output parts. The reduced width waveguides are formed by placing septa in normal size waveguide. An S-band filter was developed using this principle; the pass-band VSWR was 1.27:1 and the insertion loss about 0.5 dB in the 2.7-2.96 GHz band. The attenuation in the stop band was as shown for the TE_{10} mode [12].

There are several other absorptive type filters that have been developed experimentally but are not presently available. These include the "trapped-mode" filter and a complementary band-stop, band-pass filter [13]. Most of this work has been done at Stanford Research Institute. The advantage of the latter design is that it results in a stop band free of additional pass bands, with attenuation increasing with increasing frequency. This should enable a reduction in size and weight over the leaky wall combination discussed above while retaining the same level of performance.

It may be added here that directional couplers can be used with reflective filters to isolate the undesired reflected energy, which can then be dissipated [14].

VHF Radar Filters

Up to this point, the discussion has concerned waveguide filters. One of the most prevalent out-of-band emission problems in regard to the limits of the standard is due to P-band radars (approximately 200 MHz and 400 MHz) that do not use waveguide transmission line. There are three filter types that could be developed to provide suitable configurations from the standpoint of operational requirements. The first is the multi-resonator filter, the second is the coax leaky wall filter, and the third is the coax low-pass filter.

A four resonator VHF (300 MHz) multi-resonator filter was built by SRI [15]. Its characteristics are shown in Figure 11.5. This is a band-pass type built into a copper box (8 x 10 x 40 inches) and consisting of four coupled inductive stubs, each 3 inches in diameter. Each stub acts as a piston, with the circular head of the piston and the cover plate forming a variable capacitance that resonates with the stub at the operating frequency. The

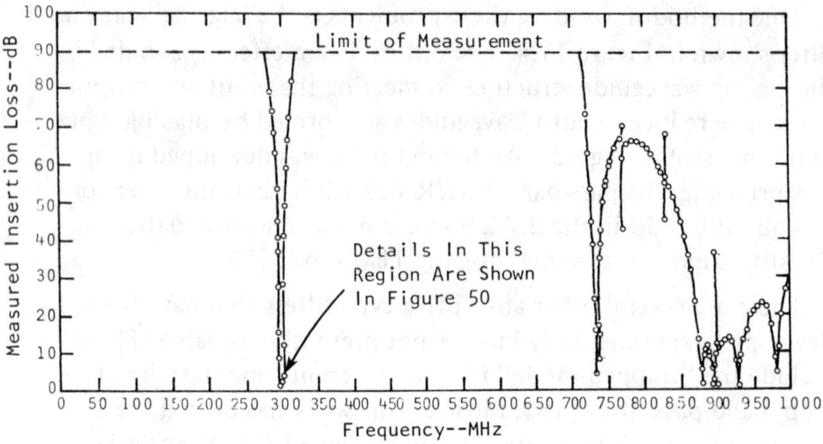

(a) Insertion loss of four-resonator VHF filter

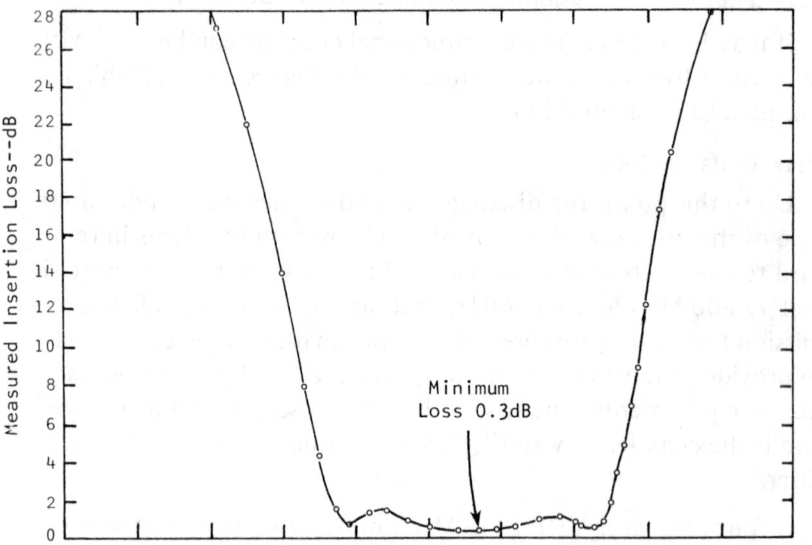

(b) Insertion-loss detail of VHF multi-resonator filter

Figure 11.5

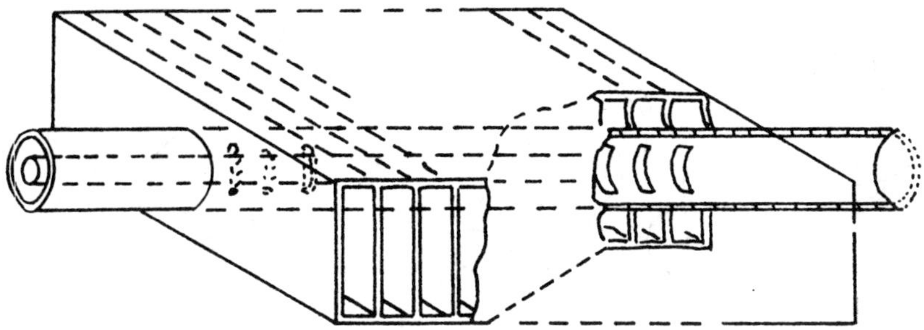

Figure 11.6 A coaxial leaky-wave filter

coupling between adjacent stubs is controlled by the placement of pairs of 3/4 inch rods that form a connection between the base and the cover plate.

The major problem anticipated in applying this filter to high power radars will be the power handling capability. It will probably be necessary to use a type of vacuum capacitor termination. Note the drop in attenuation as the frequency increases. This type of filter, if developed, will therefore only be useful for attenuating close-in spurious. Harmonics would require the additional use of the coax leaky wall filter, described below.

The coax leaky wall filter shown in Figure 11.6 is analogous to the waveguide leaky-wall filter, both in construction and theory. Energy above the cutoff frequency of the secondary waveguides is propagated to absorbing material. This filter would therefore complement the operation of the multi-resonator filter above, or could be used separately when only the harmonics are of interest.

A low-pass filter that should be able to attenuate both the above center frequency spurious and important harmonics of VHF radars is the high-power coaxial low-pass filter [16]. This filter can be built either as a reflective or absorptive type and offers the advantage of stop bands below 1000 MHz without requiring large or bulky structures. The physical appearance of this filter is illustrated in Figure 11.7. Note that it is basically a section of coax with filter elements affixed to the center conductor. The specifications of two such filters built for application to cw systems are contained in Table 11.5. Note that the listed power levels are con-

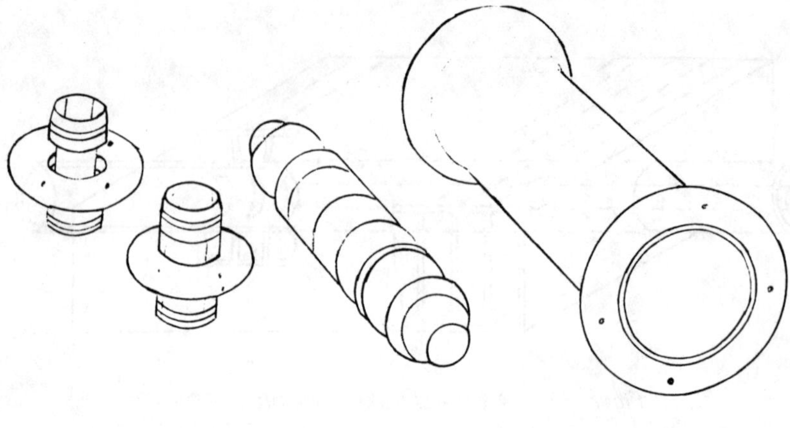

Figure 11.7 High power coaxial low-pass filter

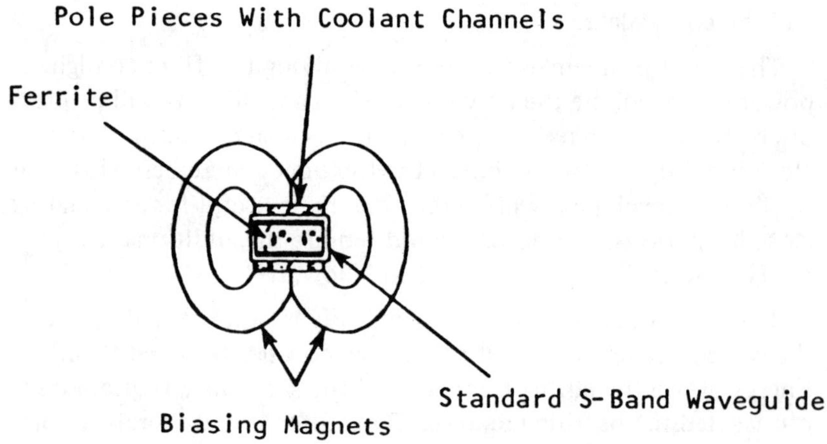

Figure 11.8 Cross section of a ferrite harmonic suppressor

siderably lower than the peak power requirements for radars (about 1 MW). The extent of this problem can be analyzed by determining the minimum gap existing between the filter elements and the outer conductor. For the filter illustrated this is about 1″ indicating that adequate power handling capability should be possible to achieve.

Another problem is the possible arcing at the power output tube due to the use of the reflective filter, which, from the viewpoint

TABLE 11.5

High Power Coaxial Low Pass Filters
Specifications

TYPES	EMT-1123	EMT-1140
Filter type	Reflective	Absorptive
PASS BAND		
Frequency Range	345 to 455 Mc	406 to 500 Mc
Insertion Loss	0.05 db	0.15 db
Input VSWR	1.25:1 Max.	1.15 Max.
Pwr trans Capacity	150 kw at 345 Mc	11 kw CW
STOP BANDS		
No. 1		
Frequency Range	690-910 Mc	812 to 1000 Mc
Attenuation	50 db	35 db
No. 2		
Frequency Range	1035-1365 Mc	1218 to 1500 Mc
Attenuation	45 db	45 db
No. 3		
Frequency Range	1380-1830 Mc	1500 to 6000 Mc
Attenuation	35 db	45 db decreasing to 10 db
Material	Copper	Aluminum
Size	20″ long/8″ diam.	9″ x 36″ x 29″
Weight	25 pounds	200 pounds
Oper. Temp/Range	−29 to 52°C	
Connector EIA Std.	6-1/8″ coax.	3-1/8″ coax.
Cooling	Air Convection	Air Convection

of size and weight, seems much more attractive than the absorptive configuration. This problem can be approached from the filter positioning aspect by mounting the filter close to the power output tube so as to prevent high potentials from building up.

Ferrite and Open Periodic Filters

Two other waveguide filter types should be mentioned, the ferrite filter and the open periodic filter. The ferrite filter is a nonreflective type that relies on the phenomenon of ferro-magnetic resonance (see Figure 11.8). However, this type of filter requires a

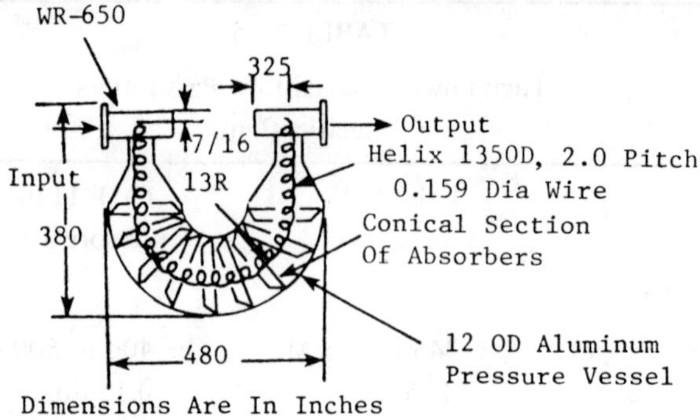

Figure 11.9 (a) L-band helix open periodic filter

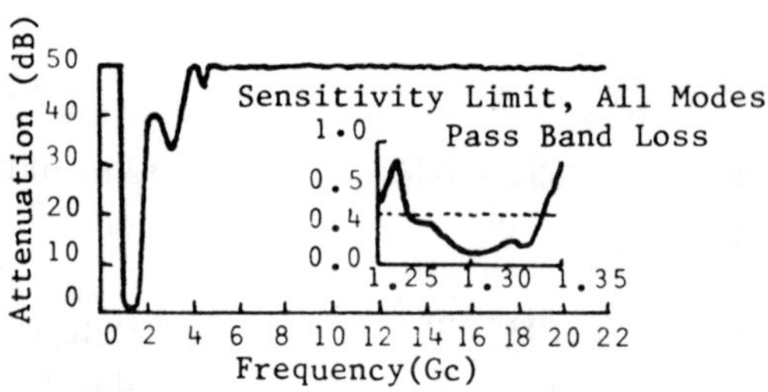

(b) Overall attenuation of the helix open periodic filter

magnetic field and cooling system for high power use, and has a relatively narrow stop band.

The open periodic filter is based on the intrinsic characteristic of an open periodic structure to propagate desired band energy from the input to the output port. The undesired energy is radiated from the structure and absorbed in surrounding radar absorbing material (Figure 11.9). The positive features of this filter include an infinitely wide stop-band, a frequency-sensitive rather than mode-sensitive characteristic, and light weight. The negative features are high insertion loss, high sensitivity to shock and vibration, and a relatively high VSWR. This filter has not been fully developed as yet. A 5 MW S-band configuration exhibited 0.8 dB insertion loss [17]. The stop band attenuation is in excess of 50 dB.

Those waveguide filters available commercially for harmonic suppression include the waffle-iron filter, the straight-through and circular leaky-wall filters, and filter sets made up of these filter elements. Trade-off curves for these filters are shown in Figures 11.10 [18] through 11.12.

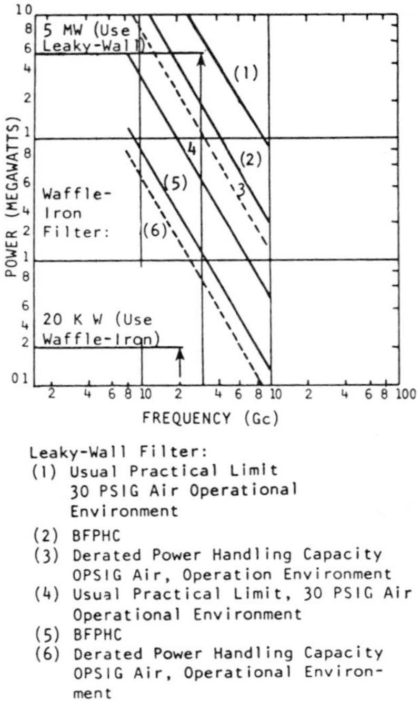

Figure 11.10 Typical power handling capacity of waffle-iron and leaky wall filters

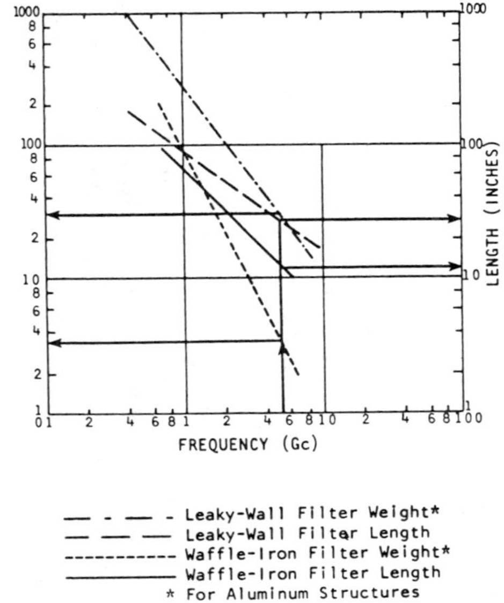

Figure 11.11 Comparison of weight and length of waffle-iron and leaky-wall filters (approximate values)

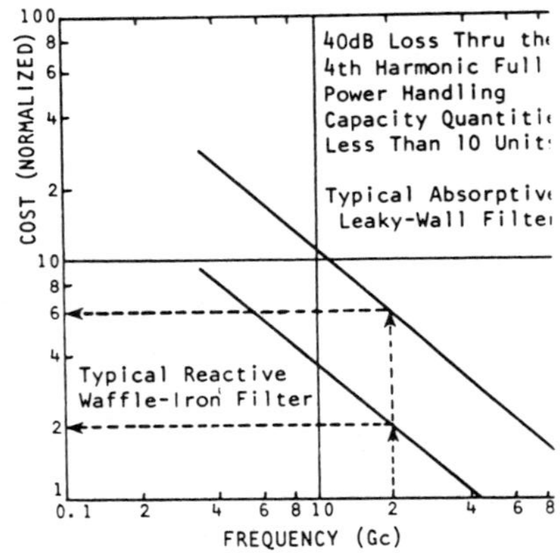

Figure 11.12 Small quantity cost comparison of waffle-iron and leaky-wall filters

SUMMARY AND CONCLUSIONS

Prior sections of this report discussed the suppression devices and techniques appropriate for the purpose of enabling high power radar transmitters to meet emission spectrum level standards. Basically, the problem areas were categorized as:

(1) In-band modulation spectrum
(2) In-band spurious emission
(3) Out-of-band spurious and harmonics.

The in-band modulation spectrum is a function of the time wave-form of the pulse, phase shift within the pulse, and near in spurious emissions. In order to meet the standard for high power values, it is necessary to shape pulses. This, in turn, requires the use of an amplifier output stage rather than an oscillator-amplifier such as a magnetron. These output amplifier devices are currently available.

The use of klystron amplifiers along with appropriate pulse shaping circuitry appears to be the best approach to both problem areas (1) and (2) at this time. The amplifiers are available commercially for the power levels of interest.

The third problem area, the suppression of out-of-band spurious and harmonics, may be alleviated in the future by the use of devices such as the axially symmetric cavity klystron but, in order to meet the limits of the standard, filters may also be required. The presently available line of commercial filters is adequate to meet the limits of most standards, but the configurations are rather heavy and bulky. Newly developed filter types that are smaller and lighter than available types have proven feasible, but require further development to be considered operationally available.

Solid state devices exhibit significantly higher spurious and harmonic levels than equivalent "hollow-state" devices, due in part to the impact noise of electrons on the lattice structure. One method that should be investigated is integrating filtering devices into stripline fabrication techniques. One approach might be deposited thin-film ferrites.

REFERENCES

[1] Yates, Ethington, et al., *Interference Reduction Techniques Study – Radar,* Technical Report ECOM-0527-F, USAECOM, Ft. Monmouth, N.J., November 1968.

[2] *Interference Notebook,* RADC-TR-66-1, Rome Air Development Center, pp. 3-44.

[3] ECAC-TN-005-89, *A Summary of the Relative Levels of Radiated Radar Harmonics,* Electromagnetic Comp. Analy. Center.

[4] MacKenzie and Everett, "A Technique of Suppressing Harmonic Emissions from High-Power Klystrons," *IEEE Electromagnetic Compatibility Symposium Record,* 1967, pp. 303-305.

[5] Brogden, Cook, and Dowds, *A Survey of Interference Reduction Techniques,* Vol. 1 of *Communications and Radars,* RADC-TDR-62-84, June 1962.

[6] Ibid.

[7] Young, Leo, "Microwave Filters to Combat RF Interference," *IEEE Transactions on EMC,* June 1968.

[8] Jones, E.M.T., *A Three-Cavity High-Powered L-Band Filter Tunable from 1250 to 1350 Mc (sic),* Technical Note 2, Contract AF 30(602)-1998, SRI for RADC.

[9] Young, L., *The Design of Branch Guide Directional Couplers for Low and High Power Applications,* RADC-TDR-62-130.

[10] Guthart and Jones, *A High-Powered S-Band Waffle-Iron Filter,* Technical Note 4, RADC-TN-61-121.

[11] *Study and Correction of Spurious Power Output (RK 6344 Magnetron),* Raytheon Company, Report PT-778, Contract Nobsr-89518.

[12] Schiffman and Young, "A New Type of Low-Pass Filter that Attenuates by Dissipation," *IEEE Transactions on Microwave Theory and Techniques,* Vol. MTT-13, No. 5, September 1965.

[13] Young, L., "Microwave Filters to Combat RF Interference," *IEEE Transactions on Electromagnetic Compatibility,* June 1968.

[14] Young, L., "Waveguide 0-dB (sic) and 3-db (sic) Directional Couplers as Harmonic Pads," *Microwave Journal,* March 1964.

[15] Young, L., E.G. Cristal et al., *Techniques for the Suppression of Spurious Energy,* RADC-TR-62-164, 28 February 1962.

[16] Special and High Power Microwave Components, Electromagnetic Technology Corporation (EMTECH).

[17] Based on conversation with Mr. John Rooney of Varian Associates.

[18] McCabe and Stone, Op. Cit.

Figure 11.4 Young, L., "Microwave Filters to Combat RF Interference," *IEEE Trans. of EMC,* June 1968. © 1968 IEEE.

Figure 11.5 Young, L., et al., Techniques for the Suppression of Spurious Energy, RADC-TR-62-164, February 28, 1962.

Figure 11.7 Special and High Power Microwave Components, Electromagnetic Technology Corporation (EMTECH).

Figures 11.9 to 11.12 McCabe and Stone, "Survey of High Power Microwave Filters," *IEEE Trans. on Aerospace and Electronic Systems,* Vol. AES-1, Oct. 1965. © 1965 IEEE.

12

EMC and MIL-STD-469

GENERAL

As stated earlier, our primary consideration is RF interference between one system and another, rather than problems within the systems themselves; i.e., the inter-system rather than the intra-system problem. Often, the intra-system problem is that the equipments involved are spatially fixed relative to each other — such as the set of shipboard equipments aboard a single ship — and a specialized treatment of remedial techniques can therefore be employed. Where the environment is continuously changing, as in the case of ships during an amphibious operation, it is necessary to consider controlling the situation by establishing reasonable limits so as to minimize EMC problems in general, independent of the specific operational environment of the system at a given time.

The latter approach is one of the reasons for an EMC standard such as MIL-STD-469 ("Radar Engineering Design Requirements, Electromagnetic Compatibility"). The most economical and effective way of achieving compatibility is to establish limits on system performance by balancing off the expense of overstandardizing against the expense of encountering and solving problems in specific situations. Determining such a balance, and establishing meaningful limits based on it, is perhaps one of the most difficult tasks in the EMC field.

The electromagnetic environment is largely a man-made entity. It consists of electromagnetic signals generated by electronic equipments, plus the noise background generated by nature (cosmic noise, triboelectricity, etc.). The problem we will examine here is the deleterious effects of an undesired radar transmitter signal on other systems susceptible to this type of signal.

Considering the nature of this problem, it is logical to attempt to control interference at its source (in this case, the radar transmitter) in order to clean up the electromagnetic environment in general. This can be done by limiting the absolute power level as a function of frequency separation from the tuned frequency. The desired effect is to limit extraneous electromagnetic emissions without penalizing the system performance.

ANTENNA GAIN STATISTICS

The problem of radar-to-radar interference in a mobile environment is one characterized by constant change. The positions of the platforms, the look-angles of the radars, and the pulsed character of the signals combine to create a situation which can best be described on a statistical basis. For example, from the compatibility standpoint, it is not very acceptable to concern oneself with main-beam to main-beam interference when it is simple to establish that such a situation would occur much less than 1% of the time, if at all.

Many studies performed previously have verified that the statistics of antenna gain of high gain antennas are closely approximated by the log normal distribution (normal for dB values) when outside the main beam [1]. Other studies have investigated the effects of local site conditions on these distributions. As a result of these efforts, it is possible to describe such statistics for a wide variety of operational situations. Table 12.1 shows the mean and standard deviations estimated for the radars of interest on this study, i.e., interference statistics due to backlobe and sidelobe radiation. These values, while estimates, have proven to be reasonably accurate in the great majority of situations. The mutual gain statistics will be a normal distribution resulting from the convolution of the individual normal antenna distributions, i.e., $P(G_tG_r) = P(G_t)*P(G_r)$.

TABLE 12.1

Mean Gain and Standard Deviations for Typical Radar Antennas

1. *Isolated Installation (no close-in obstructions or reflecting objects):*

   mean = -15 dB/isotropic

   $\sigma = 8$ dB

2. *Average Installation:*

   mean = -10 dB/isotropic

   $\sigma = 6$ dB

3. *Congested Installation (many close-in obstructions or reflections):*

   mean = -5 dB/isotropic

   $\sigma = 4$ dB

Since the distributions of the individual antenna gains are considered to be Gaussian, then the convolution (or the statistical distribution of the mutual gain) is also Gaussian, with a mean equal to the sum of the individual means, and a standard deviation equal to the square root of the sum of the squares of the individual standard deviations.

SPECTRAL LIMITS

The interpretation of the significance of the limit of spectral level in MIL-STD-469 is a difficult task, due to the complexity of the subject itself and to semantic difficulties, and further complicated by questionable measurement procedures. The term "spectral power density", in units of power per unit frequency, e.g., dBm/kHz, is usually defined as a descriptor of the frequency spectrum of a signal, independent of any receiver. A periodic pulse signal power spectrum consists of discrete spectral lines separated by a frequency difference equal to the pulse repetition frequency. The total power over a band of frequencies would therefore be equal to the sum of the power in the individual lines.

As mentioned earlier, the following expression relates the peak power and the center-frequency spectral power density of a rectangular time waveform pulse:

$$P = 10^{-3} P_s(f_o) T/\tau^2$$

where P = peak power measured in mw.

$P_s(f_o)$ = spectral power density at spectrum center measured in mw/kHz.

T = period measured in seconds

τ = pulse width (time between half power points) measured in seconds.

The overall emission spectrum limits of MIL-STD-469 are contained in Table 12.2. In this section of the report, the probability of interference using these limits as the transmitter output levels will be investigated.

SPURIOUS RADIATION ANALYSIS

The entries for spurious radiation limits in Table 12.2 are implied to be in terms of spectral power density. Radar receiver sensitivity is calibrated in terms of minimum discernible signal (MDS), a quantity measured in terms of the minimum peak power design pulse the receiving system will recognize as a target. This sensitivity measure is therefore dependent on the specific time waveform of the design signal. Based on the assumptions that a postulated receiver is fairly close in bandwidth to the interfering system, that spurious and harmonic signals are discrete signals resembling the fundamental signal, and that the limit of spectral level (dBm/kHz) of MIL-STD-469 would apply to the central spectral density of the spurious emission, the received peak power at one mile can be calculated using the classical free-space relationship

$$P_r = \frac{P G_t G_r \lambda^2}{(4\pi D)^2} .$$

TABLE 12.2

MIL-STD-469 Limits

Radar Center Frequency (MHz)	Emission Bandwidth Limits* (dBm/kHz/meter^2)	Spurious Radiation Limits** (dBm/kHz)
100 to 400	-82	-42
400 to 1215	-76	-36
1215 to 2700	-69	-29
2700 to 5000	-56	-16
5000 to 8500	-50	-10
8500 to 40,000	-45	- 5

*Radar emission bandwidth – The radar emission bandwidth shall be the frequency bandwidth which includes the fundamental frequency and extends from the lowest to the highest frequencies outside of which the levels of spectral power density do not exceed those specified hereinafter. These levels are in the main beam at a reference distance of 1850 meters and may be expressed in terms of energy received on a surface area of the receiving antenna of one square meter, i.e., joules/meter^2. For the purpose of the standard, however, the more conventional dBm/kHz/meter^2, or milliwatts x kHz^{-1} x meter^{-2}, is used.

Modulation	*B.W.*
Pulse	$20/\tau$
Pulse Compression	$20d/\tau$

where τ is the transmitter pulse width and d is the frequency dispersion factor.

**Applies to all emissions beyond the emission bandwidth defined below.

TABLE 12.3

False Target Interference

Frequency (MHz)	Spectral Density (dBm/kHz) Per (MIL-STD-469)	Free Space Gain	Median Antenna Gain	Peak Received Power (dBm) at 1 Nautical Mile: Search	Peak Received Power (dBm) at 1 Nautical Mile: Pulse Doppler
200	-42	- 84	-20	-94	-103
400	- 36	- 90	-20	-94	-103
1300	- 29	-100	-20	-97	-106
3000	- 16	-107	-20	-91	-100
5500	- 10	-113	-20	-91	-100
9500	- 5	-117	-20	-90	- 99

To perform these calculations, the following conditions were assumed:

Transmitter Modulation Characteristics:

(a) Pulse width, $\tau = 4$ μsec.
PRF, 1/T = 400 pps

(b) Pulse width, $\tau = 1$ μsec.
PRF, 1/T = 50,000 pps

Transmitter emission level: As defined by MIL-STD-469

Median Antenna Gain Product: −20 dB

Standard Deviation of Antenna Gain Product: 8.5 dB

Receiver Characteristics:
Bandwidth: 1 MHz
Minimum discernible signal: −110 dBm

Separation: 1850 meters

The lower prr mode transmitter is intended to represent a typical long range search radar. The higher prr mode transmitter is intended to represent a typical pulse-doppler tracker. Table 12.3 indicates typical results for a variety of possible operating frequencies.

For a radar receiver sensitivity of −110 dBm, some additional protection would be required beyond that offered by the MIL-STD-469 limits to keep the interference signal below the receiver noise level. This could be achieved by many means, e.g., off-tuning from the center of the spurious, video processing, etc.

The previous calculation represents one extreme relative to the character of the interference signal. That extreme considers that the spurious radiation source has a (sin x)/x frequency distribution, which is probably a very unlikely situation.

Another interpretation of the spurious radiation limits is possible by studying the measurement procedures used to obtain the "spectral power density" in compliance with MIL-STD-469. The procedure is to determine the power (dBm) received in a narrow-band receiver (spectrum analyzer), and subtract from this value an amount corresponding to ten times the logarithm of the receiver bandwidth (kHz) to yield the quantity referred to as "spectral power density" (dBm/kHz). This procedure implies that the criteria for radar receiver degradation corresponds to average re-

TABLE 12.4

Median Interference Level in a 1 MHz BW Receiver at One Mile (Assuming Noise-like Interference)

Frequency (MHz)	Spectral Density (dBm/kHz) per MIL-STD-469	Noise Level (dBm)
200	-42	-116
400	-36	-116
1300	-29	-119
3000	-16	-113
5500	-10	-113
9500	- 5	-112

ceived power rather than one which relates to the time-waveform resulting from the received spectrum. This would be true if the out-of-band energy is very noise-like and incoherent, and if the criterion is desensitization of the radar receiver rather than false targets appearing at the output.

If the assumption is made that spurious signals resemble bandwidth-limited white noise, and that the performance criterion of interest is desensitization of the radar receiver, then the term "power spectral density" would represent equivalent noise power per unit bandwidth, and the effect on a radar receiver could be obtained by multiplying it by the receiver bandwidth. For the previous set of conditions, the values shown in Table 12.4 are obtained.

The levels in Table 12.4 appear to be reasonable from the viewpoint of the standard protecting radar receivers from such desensitization effects. It was pointed out earlier that spurious radiations are not likely to duplicate the characteristics of the fundamental signal; neither are they likely to duplicate the characteristics of white noise. The actual level is probably somewhere between these two extremes.

Analysis of Other Radiation Outside the Acceptance Bandwidth

The emission bandwidth limitation of MIL-STD-469 ($20/\tau$ for a standard pulse radar) is probably based on the necessary bandwidth concept, i.e., the minimum value of the occupied bandwidth necessary to insure the required quality of performance for the systems

TABLE 12.5

Interference Level at $20/\tau$ in a 1 MHz BW Receiver (Assuming Noise-like Interference)

Frequency	MIL-STD-469 Level (dBm/kHz/meter^2)	Median Receiver Antenna Aperture*	Median Received Noise Power (dBm)
200	-82	17	-110
400	-76	24	-110
1300	-69	34	-113
3000	-56	41	-107
5500	-50	46	-107
9500	-45	51	-106

*Based on $A = \frac{G\lambda^2}{4\pi}$ (in dB relative to 1 m^2)

concerned. The $20/\tau$ value implies certain rise times and pulse shapes that may or may not be appropriate to every system. However, it is reasonable in this type of standard to establish a specific emission bandwidth requirement, and the $20/\tau$ reference can serve adequately for this purpose. Nevertheless, it should be noted that there is nothing sacrosanct about this point, and other references could be considered.

The question of the degree of protection offered by the application of the standard generates the same consideration as the spurious radiation question: does the signal at $20/\tau$ look coherent or noise like? If the signal resembles white noise, as is implied by the measurement procedure (employing the bandwidth correction as discussed above), the the quantity "spectral power density" in MIL-STD-469 implies a noise desensitization effect. The extent of this effect for the same situation as described in Table 12.4 is illustrated in Table 12.5. Since the standard limit is in terms of power spectral density per square meter on boresight, it is necessary to assume some main beam gain for the transmitter antenna. This will be estimated at 30 dB.

From Table 12.5 it is evident that, for a -110 dBm sensitivity receiver, there could be considerable desensitization at 1 mile and

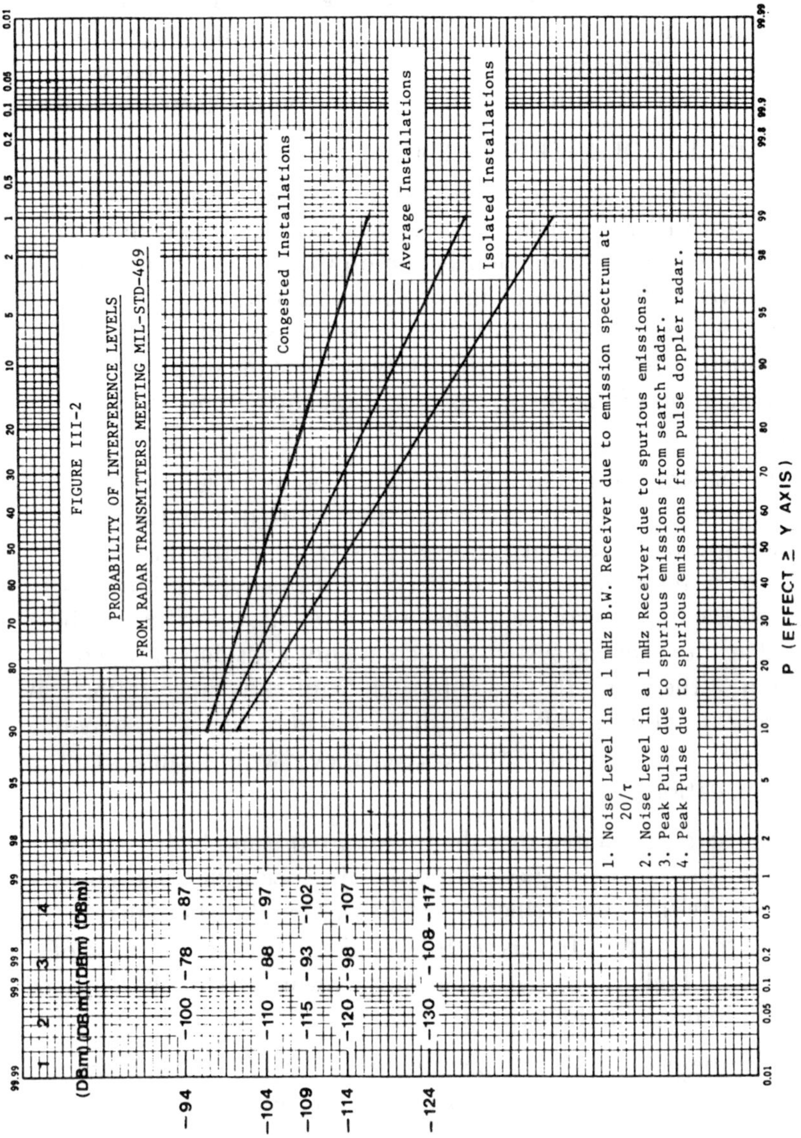

Figure 12.1 Probability of interference levels from radar transmitters meeting MIL-STD-469.

at $20/\tau$ frequency separation. As in the spurious emission case treated earlier, there is the possibility that by reconstructing a time-waveform from received frequency components, sufficient correlation between components exists to generate a pulse-like signal. The peak power of this pulse could exceed the values discussed by a considerable amount; as discussed before, such a determination would require knowledge of the pulse characteristics.

PROBABILITY OF INTERFERENCE

The statistics of the three conditions discussed above are plotted in Figure 12.1. The statistics are based on the mutual antenna gain statistics discussed previously and on the average results calculated over all frequency bands of Tables 12.3, 12.4 and 12.5. Frequency averaging was considered to be a reasonable step since the frequency dependence of the levels in the standard approximately balances off against the propagation and antenna aperture changes.

Note that for the situations discussed, there is a high probability of interference from both discrete spurious signals and from emission spectrum desensitization effects. On the other hand, it should be noted that the pulsed character of this type of interference and the non-coherence of the pulse rates allow for fixes to be applied at the receiving systems which are not considered. Inherent rejection of the interference due to such features as range gating may also occur. In other words, although undesired radar signals may be discernible at some point in the receiver they may bot cause significant degradation.

It may safely be stated that the trend of more and higher power equipment, and increasing congestion of the spectrum will eventually require that very stringent restrictions be placed on high-powered transmitters. In this regard, requirements such as those of MIL-STD-469 relative to radar transmitters will probably be imposed generally to pulse modulated systems.

One of the reasons for this observation is that, as the problem becomes more visible, other government agencies as well as the military will impose more stringent controls on transmitters as a result of the pressures of their segment of the public. Another reason is that it is believed that many problems due to RFI have been unrecognized as such, or tolerated, or both. The result is that a certain insidious degradation in performance has been accepted as the norm. The increased numbers of radar transmitters will cause this degradation unless actions are now taken to reduce this potential.

The following section will be addressed to the subject of interpreting measured data in regard to the specification limits so as to establish a basis for assessing the extent of the present situation.

REFERENCE

[1] For example: Johnson, R.C., "Mutual Gain of Radar Search Antennas," Ninth Tri-Service Conference on Electromagnetic Compatibility, October 1963.

13

EFFECTS of EMISSION SPECTRUM MEASUREMENT ERRORS

GENERAL

The purpose of this chapter is to discuss the techniques and problems in the area of measuring emission spectrum both in and out-of-band. This discussion will be limited to illustrating the possible impact of measurement errors on this study, and suggesting new methods for further investigation.
study, and suggesting new methods for further investigation.

The emission spectrum representation required for the purpose of comparing the measured data against the standard is a power vs. frequency plot over all frequencies of interest. The alternative methods used to obtain this data can be categorized as radiated vs. conducted measurements.

CONVENTIONAL RADIATED MEASUREMENTS

The conventional method for determining the radiated emission spectrum of a radar transmitter is to place a calibrated instrumentation antenna at a minimum distance of D^2/λ, where D is the largest dimension of the transmitter and λ is the wavelength of the specific emission being measured. The instrumentation antenna should be placed on the boresight of the transmitter antenna.

The determination of the radiated power is based on the relationship

$$P_t = \frac{P_r(4\pi D)^2}{G_T G_R \lambda^2} K_b$$

where P_r = received power density (power per unit bandwidth)

D = distance from transmitter to the receiver

$G_T G_R$ = mutual gain of transmitter and instrumentation antennas

λ = wavelength

K_b = bandwidth correction factor

The errors in this procedure can be categorized as follows:

1. $G_T G_R$ Errors

These errors are due to three primary causes: (1) the instrumentation antenna not on the transmitter antenna boresight at the far-field distance (D^2/λ) (for large aperture antennas this distance could be several miles especially at the higher harmonic orders); (2) the change of antenna patterns at out-of-band frequencies; and (3) errors in instrumentation antenna calibration and boresighting. The instrumentation antenna look-angle should be adjusted for maximum signal.

2. $(4\pi D)^2/\lambda^2$ Errors

These errors are due to the use of the above relationship to describe the transfer function of the energy from the transmitter antenna to the receiver antenna, when, in fact, multipath conditions exist.

3. Measurement Interpretation Errors

The interpretation of the measured data in regard to spectrum parameters is discussed in some detail in Appendix I and will not be included here.

THE EFFECT OF MEASUREMENT ERRORS ON THE DATA

The effect of not locating the measurement antenna in the far-field of the antenna under test may be estimated from the relationship between the actual gain and the far-field gain of a tapered circular aperture antenna plotted in Figure 13.1 [1]. Note that

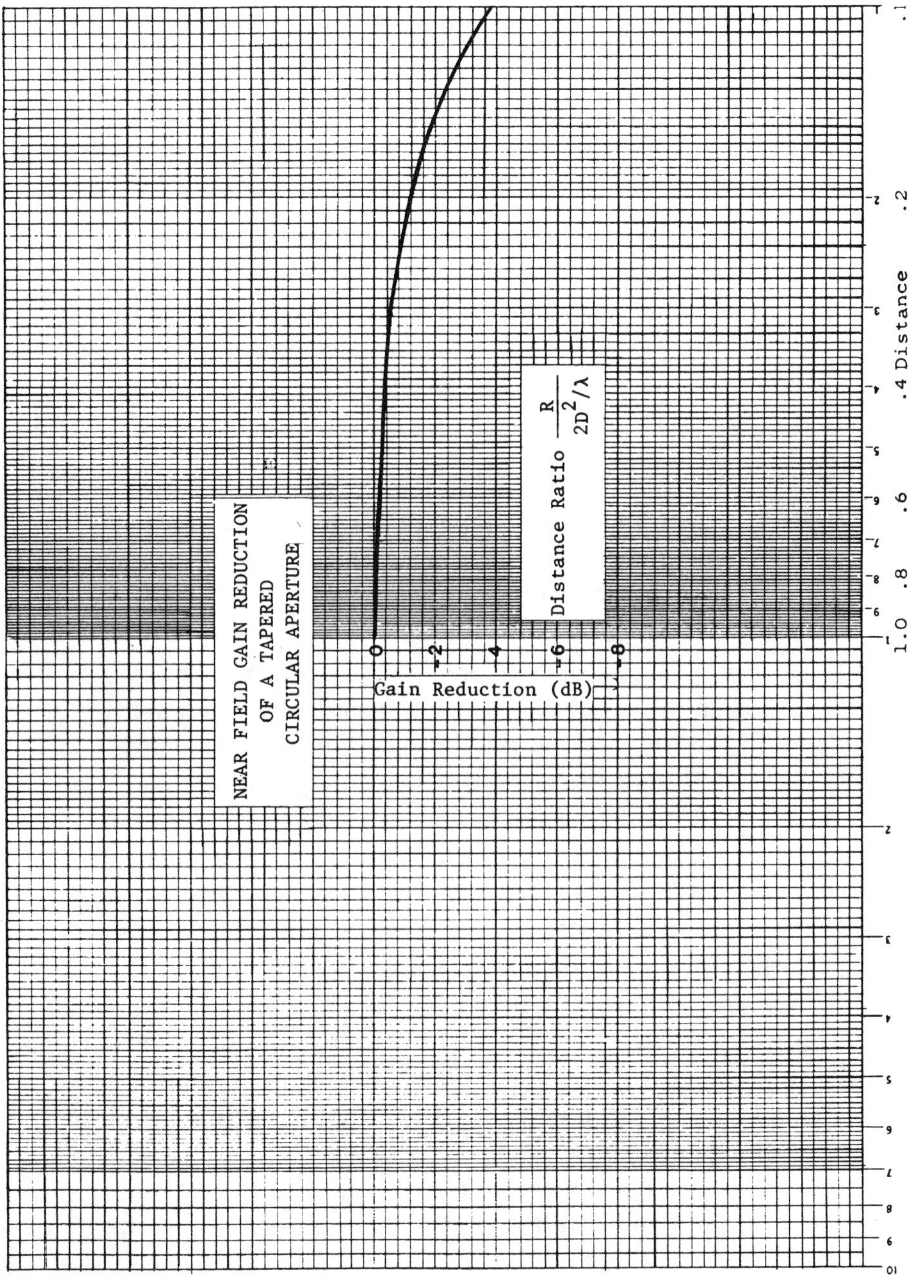

Figure 13.1 Near field gain reduction of a tapered circular aperture

TABLE 13.1
AN/FPS-8 Antenna Beamwidths
Fundamental Through 7th Harmonic

Harmonic Number	Beamwidth 3 dB	6 dB	12 dB	18 dB	24 dB
f_o	3°	4°	6°	7½°	8¾°
2	3½°	4½°	5¼°	6½°	7½°
3	3°	3½°	4°	6°	7°
4	3½°	5°	6°	7°	8°
5	3¼°	4½°	5°	5½°	7¼°
6	4°	5°	7½°	8°	10°
7	4°	5°	5°	6½°	11°

at $0.2D^2/\lambda$, there is only a 4 dB reduction in gain. Any near-field measurement will therefore tend to make the harmonic levels appear lower than they actually are, if the results are interpreted using a rated antenna gain correction figure, as is usually done.

The errors in antenna gain off-frequency are due to two primary causes, the effect of multimode propagation in the waveguide leading to the antenna, and the change in antenna dimensions in terms of units of wavelength.

The effect of the radiated pattern of a horn antenna due to multimode propagation was investigated by Salati. The significant effect is the break-up of the pattern. The results state that the gain was approximately constant up to the 7th or 8th harmonic [2].

The effect of the multimode illumination on the reflector-type antenna would be to somewhat decrease the gain on boresight, provided that the nulls are avoided. Improperly adjusting the look-angle of the instrumentation would have the same effect.

A study of such a situation [3] indicates that, based on data for the AN/FPS-8 radar, there is only a slight increase in the beamwidth for patterns up to the seventh harmonic and the splitting of the beam is quite similar to the case of the feed horn alone. This is summarized in Table 13.1.

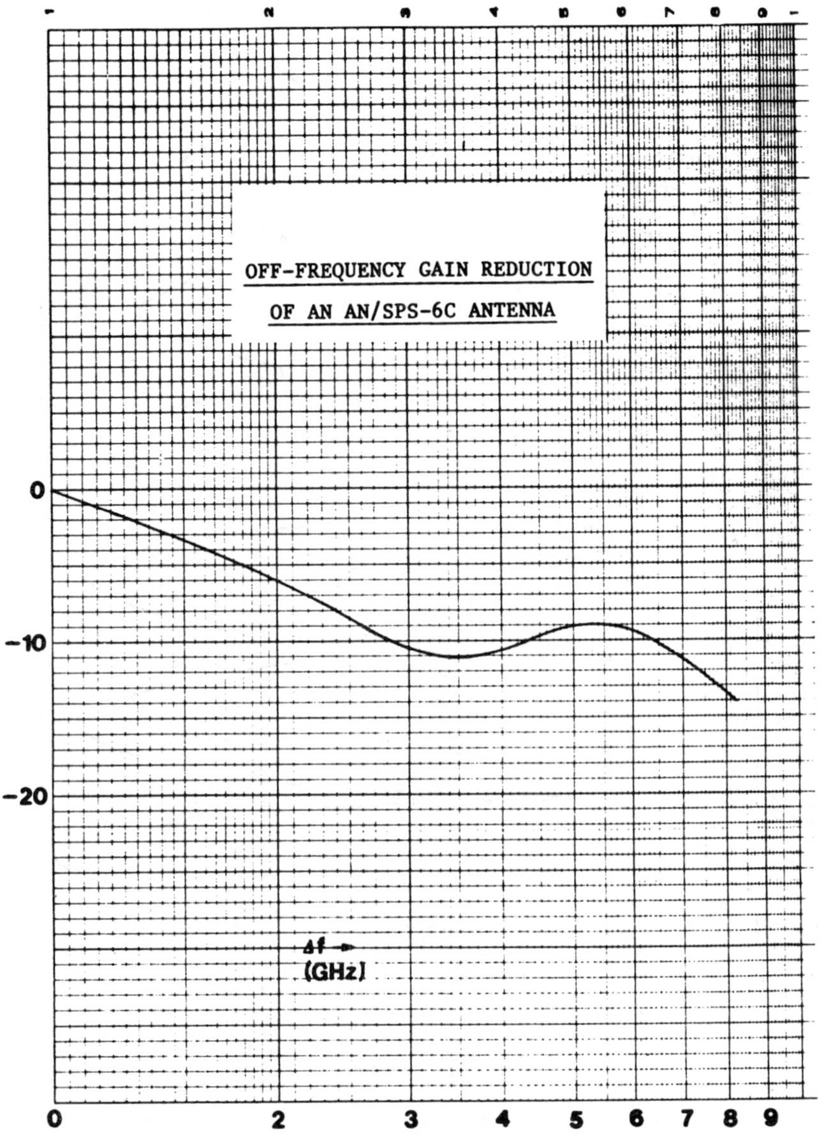

Figure 13.6 Off-frequency gain reduction of an AN/SPS-6C antenna

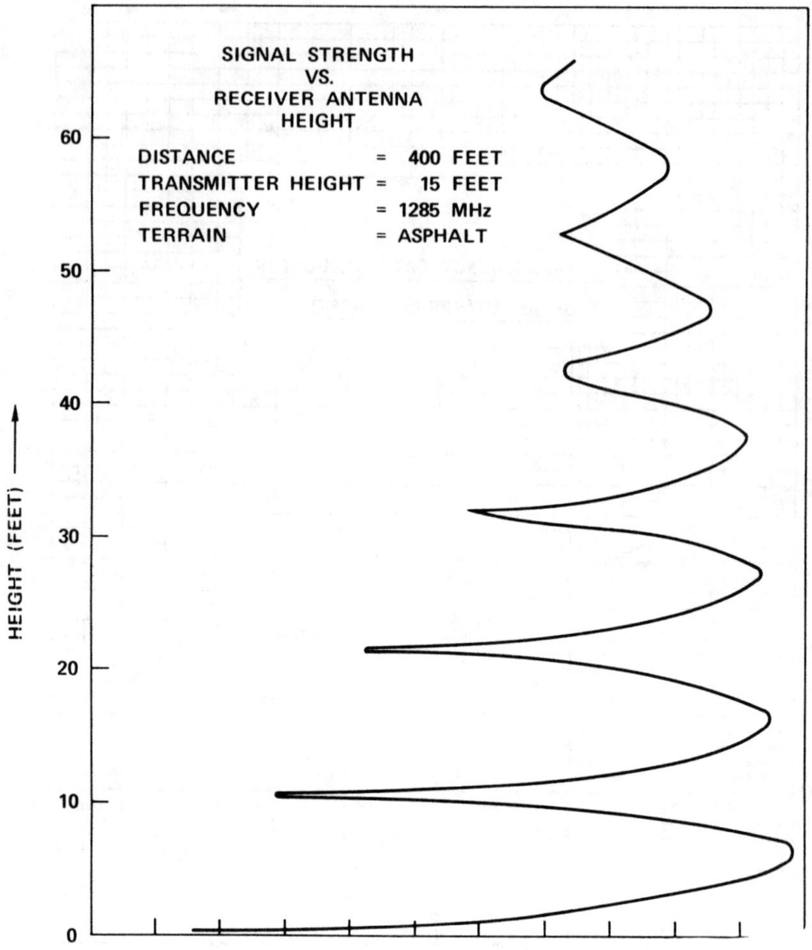

Figure 13.7 Signal strength vs. receiver antenna height

Tests performed to assess the out-of-band gain characteristics of the AN/SPS-6C antenna [4] indicate a decrease in antenna gain as the frequency increases, assuming that transmission line effects are not significant. There is not yet sufficient evidence to quantitatively judge this effect. The referenced report data for the AN/SPS-6 antenna gain is plotted in Figure 13.6.

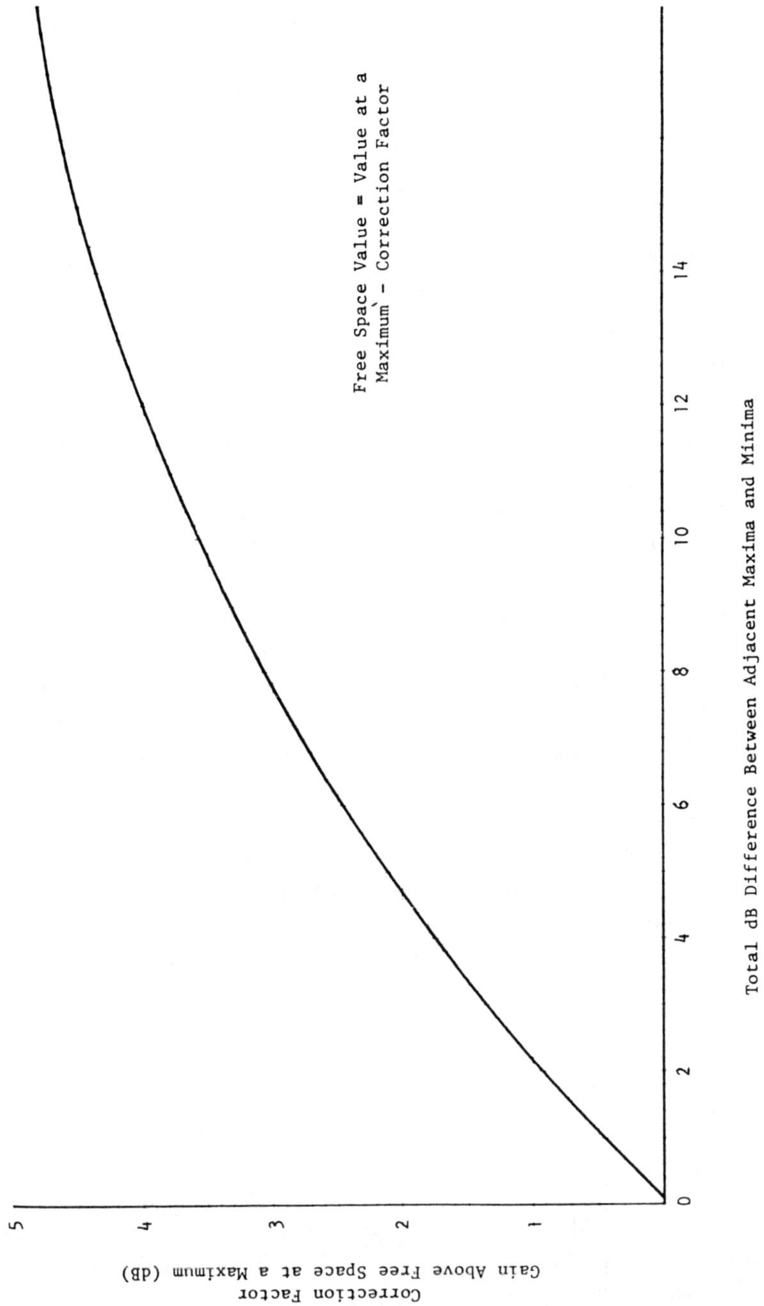

Figure 13.8 Correction curve for free space values

The effect of multipath propagation due to surface reflection is to cause a vertical lobe structure such as is shown in Figure 13.7. If there were perfect reflection from the surface there would be a maximum of 6 dB reinforcement and complete cancellation of the signal in the nulls. For less than perfect reflection, the results can be modified to allow for consideration of this effect and conversion to free space value (as shown in the correction curve in Figure 13.8). The important point here is that without the multipath effect, there is a small chance of reading signals at levels up to 6 dB higher than they actually would be, while there is a very good probability of reading them at levels lower than they would be, if the field is not probed vertically. If the field is probed over an adequate range, the errors can be corrected for.

The overall result of these errors when using the standard method of calculating radiated levels is to yield results for harmonic levels that vary from approximately 6 dB higher than is actually being radiated to theoretically infinitely lower than is being radiated (based on measuring a null). The interpretation of this error when comparing measurements against specification limits is that if the measurements indicate that the limit is exceeded by 6 dB, it is a certainty that the radiated signal is indeed above the allowed level, but if the signal does not exceed the limit there is almost no guarantee that, in actuality, it does not exceed the limit. In other words, if the conventional test method is used and the routine calculation procedure employed, and a problem is indicated, then there is an extremely high probability that this conclusion is correct. If the indication is that the level is at least 6 dB above the level for a problem situation, then the problem is certain to occur. And yet, if no problem is indicated, there is still no guarantee that a problem does not exist.

ACCURATE MEASUREMENT OF RADIATED SIGNAL LEVELS

Accurate measurement of the radiated spurious and harmonics of a high-power pulsed radar, as installed, can best be performed by following the conventional techniques, e.g., test RE-03 in MIL-STD-462 which is a detailed procedure for the conventional radiated measurement [5].

Probing of the far-field vertically is necessary in order to correct for multipath field effects as discussed above, as is properly adjusting the instrumentation antenna angle, although the referenced procedure does include a check to assure $\leqslant \pm 2$ dB tolerance using the standard correction method.

For out-of-band measurements of spurious emissions and harmonics, the problem of interpreting the results back to the antenna terminals is still unresolved, since the off-frequency behavior of the antenna is usually not available. One approach to avoiding this problem is to specify the out-of-band energy or power limits in terms of field strength or power density at various places of interest around the equipment, e.g., one mile distance and antenna boresight as is done in MIL-STD-469 for spurious emission. This combines the antenna effects with the transmitter levels, allowing for a higher confidence in comparing measured results.

IN-GUIDE EMISSION SPECTRUM MEASUREMENTS

The measurement of the in-band emission spectrum and in-band spurious emission can be performed by observing the signal through the use of a calibrated directional coupler. In-guide out-of-band measurements pose formidable problems. Directional couplers are designed to sample dominant mode energy around the center frequency. Once multi-mode conditions are reached (around the frequency of the second harmonic), the directional coupler is completely unreliable.

The in-guide measurement problem has been the subject of many studies over the years. The most significant of these for the area under discussion is the Airborne Instrument Laboratory multimode power measuring technique [6].

One of the most significant features of the equipment is a simplification of waveguide power measurement obtained by using an oversized waveguide probe section. The power flow in a standard section is described by:

$$P = -\tfrac{1}{2} R_e \frac{1}{jw\mu} \int_0^b \int_0^a \left[E_x \left(\frac{\delta Ex}{\delta z} \right)^2 + E_y \left(\frac{\delta Ey}{\delta z} \right)^2 \right] dxdy$$
$$+ \tfrac{1}{2} R_e \frac{1}{jw\mu} \int_0^b \int_0^a \left[E_x \left(\frac{\delta Ez}{\delta x} \right)^2 + E_y \left(\frac{\delta Ez}{\delta y} \right)^2 \right] dxdy$$

where a and b are the wave guide cross-section dimensions; E_x, E_y, and E_z are the complex electric field components; and x, y, and z are the coordinate system axes. The oversized waveguide approximates free space, and the power flow expression shown above reduces to

$$P = \frac{1}{2\eta}\int_0^b \int_0^a (|E_x|^2 + |E_y|^2)\,dx\,dy$$

where a and b are the oversized waveguide dimensions and η represents the impedance of free space (120π). There is a line stretcher to compensate for some of the system error. A considerable degree of automation was incorporated in the system. The accuracy for the TE_{10} mode was reported as 0.6 dB [7].

An evaluation of this device and the geometric mean probe, a similar technique, against the results obtained by integrating the power transmitted from a horn antenna over a hemisphere surrounding the equipment yielded results that showed a range of agreement between ±4 dB and ±6 dB. This seems adequate for EMC purposes, but the question of what the effects of the components up to and including the antenna are on the out-of-band emissions is still not resolved by this procedure.

SUMMARY

In summary, it appears that for all high-power pulsed equipment, i.e., above 1 kW peak power in the UHF band and above, out-of-band emission measurements should be based on the far field method at the present time. Calculations of signal level should be based on a better procedure than is presently contained in MIL-STD-469. This procedure should include a correction for multipath conditions and more consideration of the time waveform of the interference as well as its spectral characteristics.

The degree of accuracy achievable with in-guide measurement techniques appears adequate for the purpose of controlling in and out-of-band emissions. The decision on the future use of these techniques as opposed to far-field measurements should therefore be based on (1) the relative costs and availability of the measurement procedures and equipment, and (2) the resolution of the problem of the effects of the system from the point at which the measurement is performed to the point at which the energy is radiated.

REFERENCES

[1] *Microwave Engineers Handbook and Buyers Guide,* Horizon House-Microwave, Inc., December 1962, Page T-131.

[2] Salati and Lewis, "Fresnel Region and Far Field Patterns of a Horn Antenna at Fundamental and Harmonic Frequencies," *Seventh Conference on RFI and EMC,* Armour Research Foundation, November 1961.

[3] Wanbaugh, W.C., *A Summary of the Relative Levels of Radiated Radar Harmonics,* ECAC-TN-005-89, August 1967, ECAC, Annapolis, Maryland.

[4] Abromavage, M., *The Status of ECAC Special Test #9,* November 1966.

[5] *Measurement of Electromagnetic Interference Characteristics,* MIL-STD-462, 31 July 1967.

[6] *Final Report on New Methods for Measuring Spurious Emissions,* Report RADC-TDR-63-80, Feb. 1963, AD 402336.

[7] RADC-TDR-72-477, *Fifth Quarterly Progress Report on New Methods for Measuring Spurious Emissions,* September, 1962.

[8] *Correlation of In-Guide and Radiated Emission Measurement,* USAECOM, Ft. Monmouth, N.J., ECOM-01458-3, May 1966.

APPENDIX

CALIBRATION OF SPECTRUM SIGNATURES FOR DETERMINATION OF RADAR EMISSION BANDWIDTH

A. INTRODUCTION

Radar emission bandwidth is defined in MIL-STD-469, Section 3.1.14 as that frequency bandwidth, including the fundamental frequency, outside of which the power density spectrum in the main beam, at a reference distance of 1850 meters does not exceed a specified level. The level specified depends upon the frequency range in which the fundamental lies. For example, as discussed previously, for a fundamental frequency of 100 MHz to 400 MHz, the specified level is 6.31×10^{-9} mw/kHz/m^2, which is -82 dB with respect to 1 mw/kHz/m^2, commonly written -82 dBm/kHz/m^2.

Maximum allowable radar emission bandwidth is defined in MIL-STD-469, Section 6.3 according to the type of modulation used. The first three of the types of modulation listed (pulse, pulse doppler, and modified pulse) produce pulsed emissions. The remaining two types of modulation listed (CW and FM/CW) produce non-pulsed emissions. In order to determine if a particular radar complies with the maximum allowable radar emission bandwidth requirement of MIL-STD-469, it is necessary to calibrate its spectrum signature in the absolute units used in that section. Considering (1) the greater importance of radars that produce pulsed emissions, (2) the confusion that presently exists concerning spec-

---

*The author wishes to acknowledge Mr. Kitt Gillilland for his help in the development of the Appendix.

trum signature calibration of such radars, and (3) the study time available, discussion in Appendix I will be limited to such radars.

In order to develop a technique for calibrating spectrum signatures in the absolute units used in MIL-STD-469, it is necessary to first define the spectrum signature test technique assumed to have been used in obtaining the spectrum signatures. The definition should not be too restrictive, since such a definition would eliminate too many spectrum signatures already made. The definition must be exact enough, though, to allow proof of the validity of a calibration technique. Such a definition is provided in Appendix I-B which immediately follows.

Once the spectrum signature test technique is defined, it is possible to derive the desired calibration technique by making some additional assumptions about the radar emission itself. The desired calibration technique is developed in Section C below.

B. ASSUMED SPECTRUM SIGNATURE TEST TECHNIQUE

Throughout Appendix I, it will be assumed that the spectrum signature test technique used conforms to that required by MIL-STD-469, Sections 30.4-30.4.3.7. This test technique is described below. Also, throughout Appendix I, it will be assumed that the spectrum signature test technique used meets certain additional conditions not mentioned in MIL-STD-469. These conditions are given in B.2 below.

1. MIL-STD-469 Requirements for Spectrum Signature Test Technique

MIL-STD-469, Section 30.4.3, requires an open field spectrum signature test to determine the radar emission bandwidth. A minimum test site distance, as defined in Section 3.1.2 of the standard, is required. The minimum test-site distance is equal to D^2/λ or 3λ, whichever is larger, where D is the maximum aperture dimension of the largest antenna, and λ is the wavelength at the fundamental frequency. If the test antenna aperture (D_2) is larger than one-tenth of the aperture (D_1) of the antenna being measured, then the minimum test-site distance is $(D_1 + D_2)^2/2\lambda$. Section 30.4.3 of the standard also gives the test set-up block diagrams, reproduced in Figures A1-1 and AI-2.

MIL-STD-469, Sections 30.4.3.1-30.4.3.7 require the following test procedure:

(1) Tune the radar to a standard test frequency and adjust its output to the nominal level. Measure and record the transmitter power at the start of this test and at intervals of 30 minutes or less. The output power shall be measured using a thermal-type power meter and an appropriate signal sampler. The system directional coupler shall be used, if available. A power meter with a recorder output is desirable to permit recording the level of the transmitter power output, providing a permanent record of any power fluctuations occurring during the test.

(2) After adjusting the radar transmitter and measuring its output power, align the system and test antennas for maximum signal transfer as follows. The test antenna shall be elevated to a height at which the electrical centers of the two antennas are aligned. Polarization of test antenna shall be the same as that of the system antenna. With the test antenna connected to the spectrum analyzer through the appropriate transmission line and calibrated attenuators, as required, tune the analyzer until the radar signal spectrum appears centered on the analyzer display. Adjust the analyzer controls and external attenuators to provide an on-scale indication. Once this is accomplished, ascertain that the system and test antennas are aligned in azimuth, elevation, and polarization for maximum signal transfer.

(3) As specified in Section 20.4 of the standard, adjust the analyzer 3 dB bandwidth (B) to less than $1/10\tau$ or $2d/10\tau$ for fixed frequency pulses or frequency deviated pulses respectively where d is the pulse compression ratio. Adjust the dispersion of the analyzer to a value corresponding to the maximum allowable radar emission bandwidth as specified in Section 6.3 of the standard and reproduced below.

(4) With all other controls set in accordance with the operating manual of the instrument, photograph the spectrum analyzer display using an oscilloscope camera.

(5) Remove attenuation from the line and re-center the display, if required. Photograph the display. Remove only enough attenuation to provide a signal of approximately 7 dB (linear analyzer

Maximum allowable radar emission bandwidth – The radar emission bandwidth employing the following types of modulation shall not exceed the limits indicated:

Type modulation	Maximum allowable Radar emission bandwidth (MHz)
Pulse	$\frac{20}{t}$
Modified Pulse (chirp, matched filter, pulse compression and pulse stretch type radars)	$\frac{20d}{t}$ or 1/4 percent of f_o (whichever is greater)
Pulse Doppler	$\frac{20}{t}$
CW	$3 \times 10^{-4} f_o$
FM/CW	$2\Delta f + 3 \times 10^{-4} f_o$

display) or 10 dB (logarithmic display) above the noise at the portions of the display corresponding to the end frequencies of the radar emission bandwidth.

(6) Connect a calibrated signal generator to the spectrum analyzer and, with the generator set for a CW output, tune the generator to the center of the display with an amplitude at the top horizontal graticule line. Record the signal generator output level. Repeat this procedure for each horizontal line. Measure and record the signal generator frequency using the following procedure as specified in Section 10.4.2-10.4.3. In cases where the frequency to be measured is within the limits of the transfer oscillator or frequency counter, the procedure for determining frequency shall be as described in the operating instructions for the instrument. For frequencies above the limits of the transfer oscillator, the following procedure shall be used, referring to the block diagram of Figure AI-3, which was reproduced from Section 10.4.2.1 of the standard:

(a) Mix the signal source frequency with a harmonic from a signal generator whose fundamental frequency lies within the fre-

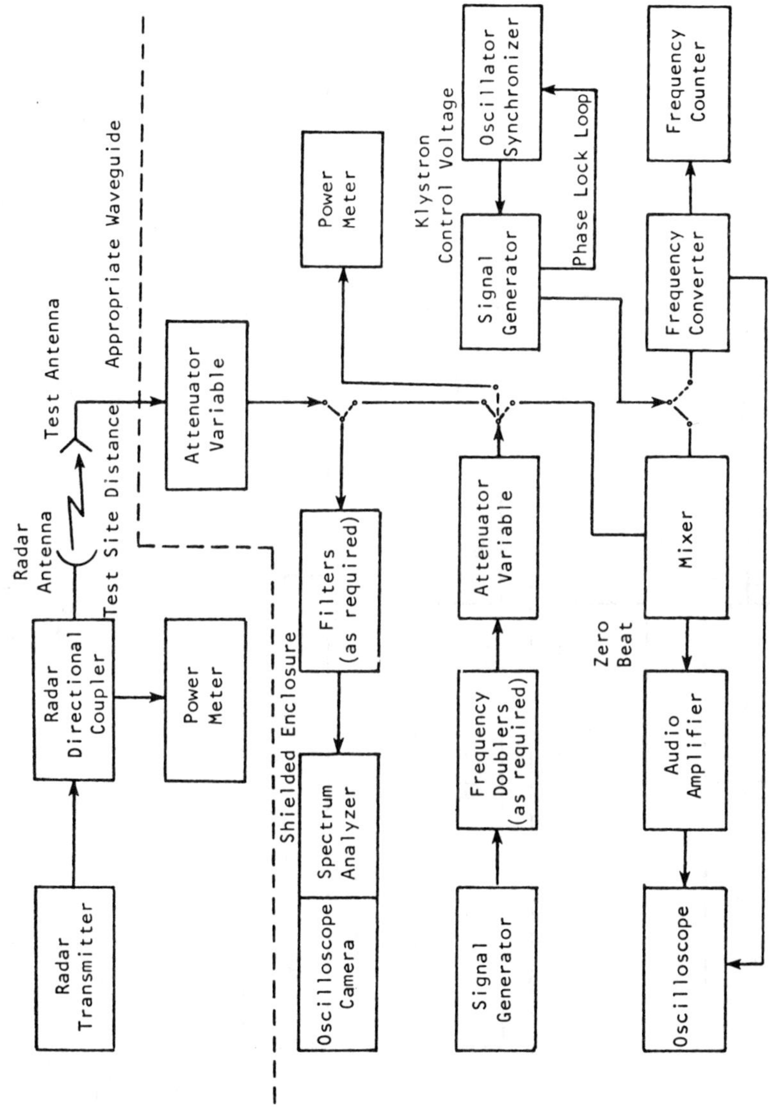

Figure AI-1 Radar emission bandwidth measurement block diagram (below 10 GHz).

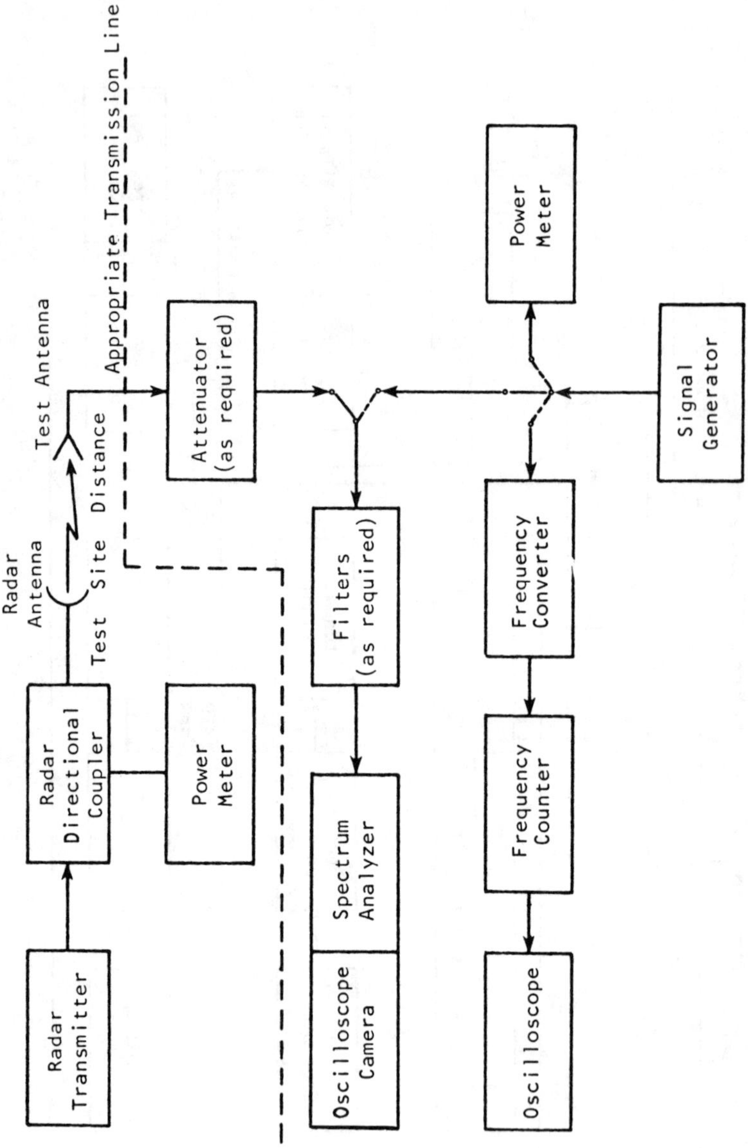

Figure AI-2 Radar emission bandwidth measurement block diagram (10 to 40 GHz).

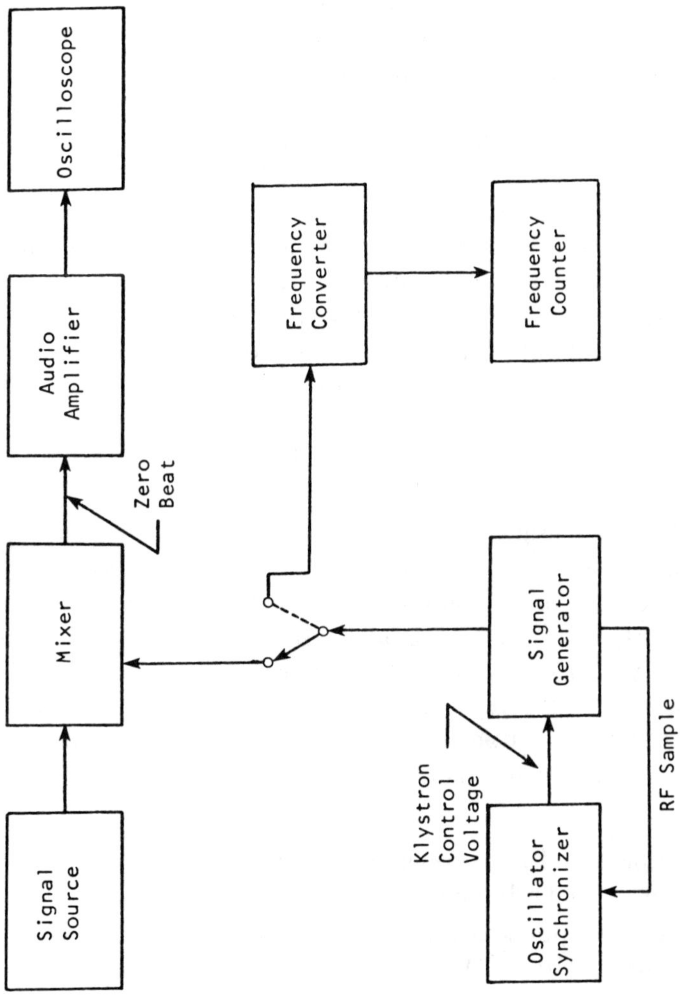

Figure AI-3 Frequency measuring equipment block diagram (10.0 to 40 GHz)

quency limits of the transfer oscillator/frequency converter. Feed the mixer output to an audio amplifier to provide sufficient signal level to be observed on an oscilloscope. Vary the frequency of the signal generator until one of its harmonics (predetermined) mixes with the signal source, providing a zero beat as observed on the oscilloscope.

(b) Switch the output of the signal generator to the transfer oscillator/frequency converter and determine the frequency of its fundamental signal. This value multiplied by the appropriate harmonic number will provide the frequency of the signal source. If an audio amplifier is unavailable for the zero beat technique of (a), then an alternate method which may be used is to replace the amplifier by an FSVM (e.g., an NF-105) and tune it to a specific frequency (e.g., 100 MHz). By adjusting the frequency of the signal generator for a peak indication on the FSVM, the frequency of the signal source may be determined by measuring the signal generator frequency above. Care must be taken to assure that the difference frequency, as set up on the FSVM, is added to or subtracted from the signal generator harmonic frequency as appropriate.

(7) Retune the signal generator to the center of the analyzer display and adjust the output level for an amplitude corresponding to the center horizontal line. Tune the signal generator frequency so that the display is coincident with the right-hand end of the analyzer display. Increase or decrease the signal generator output as required to regain a mid-scale vertical deflection. Record the generator output and frequency at this point, then tune the generator to the left-hand end of the display and repeat the process.

2. Additional Conditions Assumed Met by the Spectrum Signature Test Technique

Throughout Appendix I, the following additional conditions, not specified in MIL-STD-469, will be assumed to be met by the spectrum signature test technique used.

(a) $T_{sweep} \geqslant F_s/B^2$ where T_{sweep} is the horizontal sweep period of the spectrum analyzer, B is the spectrum analyzer IF amplifier 3 dB bandwidth, and F_s is the range of RF frequencies swept during one horizontal sweep of the spectrum analyzer. This condition assures that there is no significant loss in sensitivity due to a high sweep rate, and is usually routinely observed by spectrum analyzer operators.

(b) $1/PRF - \tau \gg 1/B$ where PRF is the pulse repetition frequency, τ is the pulse duration (time between half power points), and B is the spectrum analyzer IF amplifier 3 dB bandwidth. This condition assures that any transients occurring in the spectrum signature test set up, as a result of an incoming radar pulse, will die out before the occurrence of the next pulse.

(c) The horizontal sweep of the spectrum analyzer is not synchronized to or harmonically related to the pulse repetition frequency. This condition assures an evenly exposed spectrum signature, and is usually routinely observed by spectrum analyzer operators.

(d) The Fourier transform of the test site antenna output voltage (or field intensity*) in the frequency region of concern is (except for a multiplicative constant, possibly a constant phase shift, and possibly a zero level linear phase shift**) the same function of frequency as the Fourier transform of the test site antenna input field intensity. This condition is usually met since antennas are wide band devices.

(e) Each mixing of the incoming signal voltage (or field intensity) with a local oscillator sinusoid, for the purpose of reducing the center frequency of the incoming signal voltage (or field intensity*) is accomplished in such a manner that, within the frequency region of concern, the Fourier transform of the resultant signal voltage (after filtering) is (except for a real multiplicative constant, a constant phase shift, and possibly a zero level linear phase shift) the same function of the difference frequency as the Fourier transform of the mixer input signal voltage (or field intensity†) is of frequency. This condition is usually met closely enough, as a result of spectrum analyzer manufacturers' judicious picking of local oscillator frequencies and careful design of mixers and mixer bandpass filters.

(f) Each pre-IF amplification of the incoming signal voltage is accomplished in such a manner that, within the frequency region of concern, the Fourier transform of the output signal voltage is (except for a real multiplicative constant, a constant phase shift, and a zero level linear phase shift) the same function of frequency as the Fourier transform of the amplifier input voltage. This condition is usually met closely enough, as a result of spectrum analyzer manufacturers' design of wideband amplifiers with extremely constant voltage gains throughout the frequency range of concern.

(g) The spectrum analyzer IF amplifier unit impulse response, h(t), can be approximated as $h(t) = h_{BB}(t) \cos(2\pi f_{IF} t)$ where $h_{BB}(t)$ contains only significant frequency components that are enough lower in frequency than the IF center frequency (f_{IF}), that the IF power gain (as a function of frequency) drops to a negligible level outside the frequency region defined by

$$f_{IF} - \frac{B'}{2} \leqslant f \leqslant f_{IF} + \frac{B'}{2},$$

where B', the bandwidth of the frequency region of significant IF power gain, is related to the IF center frequency (f_{IF}) by the inequality $B' \ll f_{IF}$.

(h) The spectrum signature records a vertical deflection which is proportional to the logarithm of the spectrum analyzer peak detected IF output voltage. This mode of operation is preferable so that a clear presentation of low level side lobes is obtained, and is usually used by spectrum analyzer operators.

(i) Included on the spectrum signature is a dB scale which gives the power associated with the spectrum analyzer peak detected IF output voltage in dB with respect to some arbitrary reference.

C. DEVELOPMENT OF SPECTRUM SIGNATURE CALIBRATION TECHNIQUES

1. Introduction

In the case of radars with pulsed type emissions (pulse, pulse doppler, and modified pulse), the spectrum analyzer IF amplifier output voltage is a series of transients, one for each pulse. As a result of condition (b) of B.2, each transient dies out before the next transient occurs. According to condition (h) of B.2, a vertical deflection is produced which is proportional to the logarithm of these peak detected transients. According to condition (a) of B.2,

$$T_{sweep} \geqslant \frac{F_s}{B^2} \qquad \text{(AI-1)}$$

where T_{sweep} is the horizontal sweep period of the spectrum analyzer, B is the spectrum analyzer 3 dB bandwidth, and F_s is the range of RF frequencies swept during one horizontal sweep of the spectrum analyzer. According to step (3) of the test procedure presented in B.1,

$$B < \frac{1}{10\tau} \qquad \text{(AI-2)}$$

for fixed frequency pulses, where τ is the pulse duration (time between half power points, and

$$B < \frac{2d}{10\tau} \qquad \text{(AI-3)}$$

for frequency deviated pulses, where d is the pulse dispersion ratio. Also, according to step (3) of the test procedure presented in B.1,

$$F_s = \frac{20}{\tau} \qquad \text{(AI-4)}$$

for fixed frequency pulses, and

$$F_s = \frac{20d}{\tau} \qquad \text{(AI-5)}$$

for variable frequency pulses. Therefore, from equations (AI-1), (AI-2), and (AI-4),

$$T_{sweep} > \frac{200}{B} \qquad \text{(AI-6)}$$

for fixed frequency pulses, and from equations (AI-1), (AI-3), and (AI-5),

$$T_{sweep} > \frac{100}{B} \tag{AI-7}$$

for frequency deviated pulses. Thus for any type pulses,

$$T_{sweep} \gg \frac{1}{B} \tag{AI-8}$$

Therefore, during each vertical deflection due to an IF output transient, the horizontal deflection is negligible. Thus, only the maximum reached by the peak detected IF output voltage is recorded during each transient.

Thus, in order to derive a technique of calibrating, in the desired units, the spectrum signatures of radars with pulsed type emission, it is necessary to analyze these spectrum analyzer IF output transients in detail, and determine how they are related to the power density spectrum at the test site antenna. Since, as mentioned above, each transient dies out before the next radar pulse occurs, then the relationship between a single IF output transient and the energy density spectrum at the test site antenna, due to a single radar pulse, will be sought before dealing with power density spectra. Since the energy density spectrum at the test site antenna is directly related to the corresponding energy spectrum at the spectrum analyzer IF amplifier input, then the relationship between a single IF output transient and the energy spectrum at the IF input will be sought before dealing with the corresponding energy density spectrum at the test site antenna.

2. Relationship Between the Substituted CW Power and the Power Density Spectrum at the Test Site Antenna (Under One Assumption About the Radar Emission)

(a) Relationship between the IF Amplifier Output Transient and the IF Amplifier Input Energy Spectrum (Under One Assumption About the IF Amplifier Input Energy Spectrum)

Parseval's formula states

$$\int_{-\infty}^{\infty} |f(t)|^2 dt = \int_{-\infty}^{\infty} |F(f)|^2 df \qquad \text{(AI-9)}$$

where f(t) is any function (real or complex) which possesses a Fourier transform, F(f) given by

$$F(f) = \int_{-\infty}^{\infty} f(t) e^{-j2\pi ft} dt \qquad \text{(AI-10)}$$

Therefore,

$$\int_{-\infty}^{\infty} v_{out}^2(t) dt = \int_{-\infty}^{\infty} |V_{out}(f)|^2 df \qquad \text{(AI-11)}$$

where $v_{out}(t)$ is the spectrum analyzer IF amplifier output voltage and $V_{out}(f)$ is its Fourier transform. But,

$$|V_{out}(f)|^2 = [|H(f)|\,|V_{in}(f)|]^2 = |H(f)|^2 |V_{in}(f)|^2 \qquad \text{(AI-12)}$$

where H(f) is the spectrum analyzer IF amplifier frequency characteristic and $V_{in}(f)$ is the Fourier transform of the spectrum analyzer IF amplifier input voltage. Combining equations (AI-11) and (AI-12), dividing by the spectrum analyzer IF amplifier output resistance (R_{out}) and introducing the spectrum analyzer IF amplifier input resistance (R_{in}), the following equation is obtained,

$$p(t)\, dt = \frac{v_{out}^2(t)}{R_{out}} dt = \frac{R_{in}}{R_{out}} |H(f)|^2 \frac{|V_{in}(f)|^2}{R_{in}} df \qquad \text{(AI-13)}$$

where p(t) is the spectrum analyzer IF amplifier output power. Since the unit impulse response of the spectrum analyzer IF amplifier and the voltage input to that amplifier are real functions of time then

$$H^*(-f) = H(f) \qquad \text{(AI-14)}$$

and

$$V_{in}^*(-f) = V_{in}(f) \qquad \text{(AI-15)}$$

Thus $|H(f)|^2$ and $|V_{in}(f)|^2$ are even functions of f, and equation (AI-13) yields

$$\int_{-\infty}^{\infty} p(t)\,dt = \int_{0}^{\infty} G(f)\,E_{IF}(f)\,df \qquad \text{(AI-16)}$$

where $G(f) = (R_{in}/R_{out})|H(f)|^2$ is the power gain (as a function of frequency) of the spectrum analyzer IF amplifier, and $E_{IF}(f) = 2|V_{in}(f)|^2/R_{in}$ is the energy spectrum at the spectrum analyzer IF amplifier output. Assume $E_{IF}(f) \approx E_{IF}(f_{IF})$ for all frequencies (f) in the frequency region of significant IF power gain, defined in condition (g) of B.2 by the relation

$$f_{IF} - \frac{B'}{2} \leqslant f \leqslant f_{IF} + \frac{B'}{2},$$

where B', the bandwidth of the frequency region of significant IF power gain, is given by the inequality, $B' \ll f_{IF}$. As a result of this assumption, equation (AI-16) can be written as follows,

$$\int_{-\infty}^{\infty} p(t)\,dt = E_{IF}(f_{IF}) \int_{f_{IF}-(B'/2)}^{f_{IF}+(B'/2)} G(f)\,df \qquad \text{(AI-17)}$$

Thus,

$$\int_{-\infty}^{\infty} p(t)\,dt = E_{IF}(f_{IF})\,G(f_{IF})\,B_{eff} \qquad \text{(AI-18)}$$

where the $G(f_{IF})$ is the spectrum analyzer IF amplifier gain at the center frequency (f_{IF}), and B_{eff} is the spectrum analyzer IF amplifier effective random noise bandwidth defined as

$$B_{eff} = \frac{1}{G(f_{IF})} \int_{0}^{\infty} G(f)\,df \qquad \text{(AI-19)}$$

If a means were available to experimentally obtain the integral on the left hand side of equation (AI-18), then $E_{IF}(f_{IF})$ could be obtained directly from that equation since $G(f_{IF})$ and B_{eff} are obtainable experimentally. That integral is simply the total energy out of the spectrum analyzer IF amplifier during the transient due to a single radar pulse. As explained in C.1, the spectrum signature actually records only the maximum value of the power associated with the peak detected IF output voltage (measured in dB with respect to some arbitrary reference) during the transient. Equation (AI-18) can be rewritten in terms of this maximum value and solved for $E(f_{IF})$, the energy spectrum at the spectrum analyzer IF amplifier input evaluated at the IF center frequency (f_{IF}). The resulting expression is

$$E_{IF}(f_{IF}) = \frac{p_{pk}}{2B_tB_{eff}G(f_{IF})} \tag{AI-20}$$

where p_{pk} is the maximum value of the spectrum analyzer IF output power during the transient, which is equal to the power associated with the maximum value of the peak detected IF output voltage during the transient, and B_t, which will be called the temporal bandwidth, is defined by the equation,

$$\frac{1}{B_t} = \frac{2}{p_{pk}}\int_{-\infty}^{\infty} p(t)\,dt \approx \frac{1}{P_{env\ pk}}\int_{-\infty}^{\infty} P_{env}(t)\,dt \tag{AI-21}$$

where $P_{env}(t)$ is the envelope of $p(t)$ and $p_{env\ pk}$ is the maximum value of $p_{env}(t)$ during the transient. It should be noted that, until an additional assumption is made (in C.3) about the radar emission, the temporal bandwidth (B_t) is not a system constant, and will, in general, depend upon $E_{IF}(f_{IF})$.

(b) Relationship Between the Energy Density Spectrum at the Test Site Antenna and the Energy Spectrum at the IF Amplifier Input

As a result of conditions (e) and (f) of B.2, the energy spectrum level, $E_{RF}(f_{RF})$, at the test site antenna output is related to the energy spectrum level, $E_{IF}(f_{IF})$, at the IF amplifier input by the relation

$$E_{RF}(f_{RF}) = \frac{E_{IF}(f_{IF})}{G_cG_1G_{RF}}\,. \tag{AI-22}$$

G_{RF} is the power gain between the spectrum analyzer input and the IF amplifier input as a direct consequence of the real multiplicative constants mentioned in those two conditions and the impedance levels involved; G_c is the power gain of the cable (or waveguide) between the test site antenna and the spectrum analyzer input; G_1 is the attenuator power gain between the test site antenna and the spectrum analyzer input; $E_{RF}(f_{RF})$ is the energy spectrum at the test site antenna output (evaluated at the RF frequency, f_{RF}, where the IF bandpass is effectively tuned), and $E_{IF}(f_{IF})$ is the energy spectrum at the IF amplifier input (evaluated at the IF center frequency, f_{IF}). As a result of condition (d) of B.2, the energy density spectrum level, $E_{SD}(f_{RF})$, at the test site antenna input is related to the energy spectrum level, $E_{RF}(f_{RF})$, at the test site antenna output by the relation

$$E_{SD}(f_{RF}) = \frac{E_{RF}(f_{RF})}{A} \qquad \text{(AI-23)}$$

where A is the effective area of the test site antenna and $E_{SD}(f_{RF})$ is the energy density spectrum at the test site antenna input (evaluated at the RF frequency where the IF bandpass is effectively tuned). Combining equations (AI-22) and (AI-23) the desired relationship is obtained,

$$E_{sD}(f_{RF}) = \frac{E_{IF}(f_{IF})}{AG_cG_1G_{RF}} \qquad \text{(AI-24)}$$

(c) Radar Emission Assumption

It is obvious from equation (AI-24) that the assumption made in obtaining equation (AI-17) is equivalent to the following radar emission assumption,

Radar Emission Assumption (1):

$E_{sD}(f) \approx E_{sD}(f_{RF})$ for all frequencies (f) in the region defined by $f_{RF} - (B'/2) \leqslant f \leqslant f_{RF} + (B'/2)$, where $E_{sD}(f)$ is the energy density spectrum at the test site antenna input due to a single radar pulse, $E_{sD}(f_{RF})$ is $E_{sD}(f)$ evaluated at the RF frequency (f_{RF}), f_{RF} is the RF frequency where the IF bandpass is effectively tuned during the pulse, and B' is the bandwidth of the frequency region of significant IF power gain.

(d) Conclusion

From equations (AI-20) and (AI-24) it can be seen that under radar emission assumption (1),

$$E_{sD}(f_{RF}) = \frac{p_{pk}}{AG_cG_1G_{RF}G(f_{IF})} \; \frac{1}{2B_tB_{eff}} \tag{AI-25}$$

In this equation, the value of p_{pk} is given by

$$p_{pk} = 2P_DAG_cG_1G_{RF}G(f_{IF}) \tag{AI-26}$$

where P_D is the average CW power density (power per unit area) of frequency (f_{RF}), which, if applied to the test site antenna input, would produce the same vertical deflection on the spectrum analyzer (when the IF bandpass is effectively tuned to f_{RF}) as that produced by the radar pulse (when the IF bandpass was effectively tuned to f_{RF}). Equation (AI-26) states that the peak power of the CW IF output is equal to twice the average power, $P_D AG_cG_1G_{RF}G(f_{IF})$, of the CW IF output.

Finally, from equations (AI-25) and (AI-26), and from the relationship between energy density spectrum and power density spectrum, it is apparent that under radar emission assumption (1),

$$P_{sD}(f_{RF}) = E_{sD}(f_{RF}) \times PRF = \frac{P_D}{B_tB_{eff}} \times PRF \tag{AI-27}$$

where $P_{sD}(f_{RF})$ is the power density spectrum (power per unit area per unit frequency) at the test site antenna input (evaluated at f_{RF}), PRF is the pulse repetition frequency, $E_{sD}(f_{RF})$ is the energy density spectrum (energy per unit area per unit frequency) at the test site antenna input (evaluated at f_{RF}), f_{RF} is the RF frequency where the IF amplifier bandpass was effectively tuned during the radar pulse, P_D is the average CW power density (power per unit area) of frequency (f_{RF}) which, if applied to the test site antenna input, would produce the same vertical deflection on the spectrum analyzer (when the IF bandpass is effectively tuned to f_{RF}) as that produced by the radar pulse (when the IF bandpass was effectively tuned to f_{RF}). B_{eff} is the spectrum analyzer IF amplifier effective random noise bandwidth defined by equation (AI-19) as

$$B_{eff} = \frac{1}{G(f_{IF})} \int_0^{\infty} G(f)\, df \qquad \text{(AI-19)}$$

where $G(f) = (R_{in}/R_{out})\, |H(f)|^2$ is the power gain (as a function of frequency) of the spectrum analyzer IF amplifier, R_{in} is the spectrum analyzer IF amplifier input resistance, R_{out} is the spectrum analyzer IF amplifier output resistance, H(f) is the spectrum analyzer IF amplifier frequency characteristic, and $G(f_{IF})$ is G(f) evaluated at the IF center frequency (f_{IF}). B_t is the spectrum analyzer IF amplifier temporal bandwidth defined by equation (AI-21) as

$$\frac{1}{B_t} = \frac{2}{p_{pk}} \int_{-\infty}^{\infty} p(t)\, dt \approx \frac{1}{p_{env\ pk}} \int_{-\infty}^{\infty} p_{env}(t)\, dt \qquad \text{(AI-2)}$$

where p(t) is the spectrum analyzer IF output power during the transient caused by a single radar pulse, p_{pk} is the maximum value of p(t) during the transient, $p_{env}(t)$ is the envelope of p(t), and $p_{env\ pk}$ is the maximum value of $p_{env}(t)$ during the transient.

The value of P_D can be determined by substituting a CW signal of average power (P_{sg}) and frequency (f_{RF}) at the spectrum analyzer input which produces the same vertical deflection on the spectrum analyzer (when the IF bandpass is effectively tuned to f_{RF}) as that produced by the radar pulse (when the IF bandpass was effectively tuned to f_{RF}), and using the relation

$$P_D = \frac{P_{sg}}{AG_cG_1} \qquad \text{(AI-28)}$$

where A is the effective area of the test site antenna, G_c is the power gain of the cable (or waveguide) between the test site antenna and the spectrum analyzer input, and G_1 is the attenuator power gain between the test site antenna and the spectrum analyzer input.

It should be noted that radar emission assumption (1) does not by itself assure that the temporal bandwidth (B_t) is independent of the power density spectrum level, $P_{sD}(f_{RF})$ which is being calibrated. It will be shown in the following section (C.3) that if one additional radar emission assumption is made, such independence is assured and the temporal bandwidth becomes a system constant. Without the additional radar emission assumption, it would in general be necessary to calibrate many points along the frequency axis of the spectrum signature, redetermining the temporal bandwidth (B_t) for each point – a very impractical procedure.

3. Relationship Between the Substituted CW Power and the Power Density Spectrum at the Test Site Antenna (Under Two Assumptions About the Radar Emission)

(a) Relationship Between the IF Amplifier Output Transient and the IF Amplifier Input Energy Spectrum (Under Two Assumptions About the IF Amplifier Input Energy Spectrum)

The Fourier transform, $V_{in}(f)$, of the signal voltage at the spectrum analyzer IF amplifier input can be expressed as

$$V_{in}(f) = |V_{in}(f)|e^{j\theta(f)} \tag{AI-29}$$

and

$$V_{in}(-f) = |V_{in}(-f)|e^{j\theta(-f)} \tag{AI-30}$$

where $\theta(f)$ is the phase angle of the Fourier transform of the signal voltage. Since the signal voltage at the spectrum analyzer IF amplifier input is real, then

$$V_{in}(f) = V_{in}^*(-f) \tag{AI-31}$$

$$|V_{in}(f)| = |V_{in}(-f)| \tag{AI-32}$$

$$\theta(f) = -\theta(-f) \tag{AI-33}$$

The Fourier transform, $V_{out}(f)$, of the spectrum analyzer IF amplifier output voltage is given by

$$V_{out}(f) = V_{in}(f)H(f) \tag{AI-34}$$

where H(f) is the frequency characteristic of the spectrum analyzer IF amplifier. According to condition (g) of B.2, the unit impulse response, h(t), of the spectrum analyzer IF amplifier can be approximated by

$$h(t) = h_{BB}(t)\cos(2\pi f_{IF} t) \tag{AI-35}$$

where $h_{BB}(t)$ contains only significant frequency components that are enough lower in frequency than the IF center frequency (f_{IF}) that the IF power gain (as a function of frequency) drops to a negligible level outside the frequency region defined by $f_{IF} - (B'/2) < f < f_{IF} + (B'/2)$, where B′, the bandwidth of the frequency region of significant IF power gain, is related to the IF center frequency (f_{IF}) by the inequality, $B' \ll f_{IF}$. Thus the frequency characteristic, H(f), of the spectrum analyzer IF amplifier is

$$H(f) = \frac{H_{BB}(f-f_{IF})}{2} + \frac{H_{BB}(f+f_{IF})}{2} \tag{AI-36}$$

where $H_{BB}(f)$ is the Fourier transform of $h_{BB}(t)$ and has significant magnitude only within the frequency region defined by $-(B'/2) \leqslant f \leqslant (B'/2)$ where B′ is as defined above. Thus $H_{BB}(f-f_{IF})$ has significant magnitude only within the frequency region defined by $f_{IF} - (B'/2) \leqslant f \leqslant f_{IF} + (B'/2)$, and $H_{BB}(f+f_{IF})$ has significant magnitude only within the frequency region defined by $-f_{IF} - (B'/2) \leqslant f \leqslant -f_{IF} + (B'/2)$. The Fourier transform, $V_{out}(f)$, can be obtained from equations (AI-34), (AI-29), and (AI-36) as

$$V_{out}(f) = |V_{in}(f)| \frac{H_{BB}(f-f_{IF})}{2} e^{j\theta(f)} + |V_{in}(f)| \frac{H_{BB}(f+f_{IF})}{2} e^{j\theta(f)} \tag{AI-37}$$

where the first term need be evaluated only in the frequency region where $H_{BB}(f-f_{IF})$ has significant magnitude and the second term need be evaluated only in the frequency region where $H_{BB}(f+f_{IF})$ has significant magnitude. Assume $|V_{in}(f)| \approx |V_{in}(f_{IF})|$ for all frequencies (f) in the frequency region of significant IF power gain, defined in condition (g) of Appendix I-B-2 by the relation $f_{IF} - (B'/2) \leqslant f \leqslant f_{IF} + (B'/2)$ where B′, the bandwidth of the frequency region of significant IF power gain, is given by the inequality $B' \ll f_{IF}$. From equation (AI-32) it is obvious that, under this assumption, $|V_{in}(f)| \approx |V_{in}(f_{IF})|$ also for all frequencies (f) in the frequency region defined by $-f_{IF} - (B'/2) \leqslant f \leqslant -f_{IF} - (B'/2)$. As a result of this assumption, equation (AI-37) can be written as

$$V_{out}(f) = |V_{in}(f_{IF})| \frac{H_{BB}(f-f_{IF})}{2} e^{j\theta(f)} + |V_{in}(f_{IF})| \frac{H_{BB}(f+f_{IF})}{2} e^{j\theta(f)} \qquad \text{(AI-38)}$$

Assume, in addition, $\theta(f) \approx C_0 + C_1 x(f-f_{IF})$ for all frequencies (f) in the same frequency region as defined in the previous assumption, where C_0 and C_1 are the coefficients of the first two terms of the Taylor series expansion of $\theta(f)$ about the point $f = f_{IF}$. Similarly, it is obvious from equation (AI-33) that, under this assumption, $\theta(f) = -C_0 + C_1 x(f+f_{IF})$ for all frequencies (−f) in the corresponding negative frequency region. As a result of this assumption, equation (AI-38) can be written as

$$V_{out}(f) = |V_{in}(f_{IF})| \left[\frac{H_{BB}(f-f_{IF})}{2} e^{j(c_0' + c_1(f)} + \frac{H_{BB}(f+f_{IF})}{2} e^{j(-c_0' + c_1 f)} \right] \qquad \text{(AI-39)}$$

where $c_0' = c_0 - c_1 f_{IF}$. Define the frequency function, V(f), such that

$$V_{out}(f) = V(f) e^{jc_1 f} \qquad \text{(AI-40)}$$

Thus

$$V(f) = |V_{in}(f_{IF})| \left[\frac{H_{BB}(f-f_{IF})}{2} e^{jc_0'} + \frac{H_{BB}(f+f_{IF})}{2} e^{-jc_0'} \right]. \qquad \text{(AI-41)}$$

The inverse Fourier transform, v(t), of the frequency function, V(f), is related to $v_{out}(t)$ by a simple time delay. This fact can be seen as follows. Using the definition of inverse Fourier transform and equation (AI-40),

$$v_{out}(t) = \int_{-\infty}^{\infty} [V(f) e^{jc_1 f}] e^{j2\pi ft} df = \int_{-\infty}^{\infty} V(f) e^{j2\pi(t+(c_1/2\pi))f} df \qquad \text{(AI-43)}$$

The last integral is recognized as $v(t+(c_1/2\pi))$, and thus,

$$v_{out}(t) = v\left[t + \frac{c_1}{2\pi}\right] \qquad \text{(AI-43)}$$

Also, the right hand side of equation (AI-41) is recognized to be expressable in the following form,

$$V(f) = |V_{in}(f_{IF})| \int_{-\infty}^{\infty} H_{BB}(\nu)\left[\tfrac{1}{2}\delta[(-\nu + f) - f_{IF}] + \tfrac{1}{2}\delta[(-\nu + f) + f_{IF}]\right] e^{j[((-\nu+f)/f_{IF})c_0']} d\nu \qquad \text{(AI-44)}$$

This expression is valid because: (1) during the performance of the indicated integration of the first term, the argument of the delta function is zero only at $\nu = f - f_{IF}$ where the exponent of e has the value jc_0', thus producing the term $\tfrac{1}{2}H_{BB}(f-f_{IF})\,e^{jc_0'}$ which is then multiplied by $|V_{in}(f_{IF})|$. (2) During the performance of the indicated integration of the second term, the argument of the delta function is zero only at $\nu = f+f_{IF}$ where the exponent e has the value of $-jc_0'$ thus producing the term $\tfrac{1}{2}H_{BB}(f+f_{IF})\,e^{-jc_0'}$ which is then multiplied by $|V_{in}(f_{IF})|$. The integral in equation (AI-44) is recognized as a frequency convolution, making it possible to express equation (AI-44) as

$$V(f) = |V_{in}(f_{IF})|[H_{BB}(f)*M(f)\,e^{j((f/f_{IF})c_0')}] \qquad \text{(AI-45)}$$

where

$$M(f) = [\tfrac{1}{2}\delta(f-f_{IF}) + \tfrac{1}{2}\delta(f+f_{IF})] \qquad \text{(AI-46)}$$

inverse Fourier transform of V(f), is given by

$$m(t) = F^{-1}\left[M(f)\right] = \cos(2\pi f_{IF}t) \qquad \text{(AI-47)}$$

The inverse Fourier transform of $M(f)\,e^{j((f/f_{IF})c_0')}$ is given by

$$F^{-1}\left[M(f)\,e^{j((f/f_{IF})c_0')}\right] = \int_{-\infty}^{\infty} M(t)\,e^{j2\pi(t+(c_0'/2\pi f_{IF}))f} df \qquad \text{(AI-48)}$$

which is recognized as $m(t+(c_0'/2\pi f_{IF}))$. Thus

$$F^{-1}\left[M(f)e^{j((f/f_{IF})c_0')}\right] = \cos 2\pi f_{IF}\left[t + \frac{c_0'}{2\pi f_{IF}}\right] \tag{AI-49}$$

In general,

$$F\left[f(t)g(t)\right] = F\left[f(t)\right] * F\left[g(t)\right] \tag{AI-50}$$

Therefore, from equations (AI-45), (AI-49), and (AI-50), v(t), the inverse Fourier transform of V(f) is given by,

$$v(t) = |V_{in}(f_{IF})|\, h_{BB}(t)\cos 2\pi f_{IF}\left[t + \frac{c_0'}{2\pi f_{IF}}\right] \tag{AI-51}$$

Finally, from equations (AI-43) and (AI-51),

$$v_{out}(t) = |V_{in}(f_{IF})| \left[h_{BB}\left(t + \frac{c_1}{2\pi}\right)\cos 2\pi f_{IF}\left(t + \frac{c_0'}{2\pi f_{IF}} + \frac{c_1}{2\pi}\right)\right] \tag{AI-52}$$

and the IF output power is

$$p_{out}(t) = \frac{v^2_{out}(t)}{R_{out}} = \frac{|V_{in}(f_{IF})|^2}{R_{out}}\left[h^2_{BB}\left(t + \frac{c_1}{2\pi}\right)\right.$$

$$\left. \times \cos^2 2\pi f_{IF}\left(t + \frac{c_0'}{2\pi f_{IF}} + \frac{c_1}{2\pi}\right)\right] \tag{AI-53}$$

where $h^2_{BB}(t+(c_1/2\pi))$ is the envelope of the square of the IF amplifier unit impulse response, advanced in time by $c_1/2\pi$, and the quantity in brackets is a sinusoid of frequency $2f_{IF}$ with lower peaks equal to zero and upper peaks equal to $h^2_{BB}(t+(c_1/2\pi))$. As a result of condition (g) of Appendix I-B-2, and equation (AI-53), the maximum output power, p_{pk}, can be approximated by

$$p_{pk} = \frac{|V_{in}(f_{IF})|^2}{R_{out}}\, h^2_{BB}\left(t_p' + \frac{c_1}{2\pi}\right) \tag{AI-54}$$

where $t = t'_p$ is the time at which $h^2_{BB}(t+(c_1/2\pi))$ reaches its maximum. This is equivalent to

$$P_{pk} = \frac{|V_{in}(f_{IF})|^2}{R_{out}} h^2_{BB}(t_p) \tag{AI-56}$$

where $t = t_p = t'_p + c_1/2\pi$ is the time at which the square of the IF amplifier unit impulse response reaches its maximum.

The temporal bandwidth can now be determined. Substituting equations (AI-53) and (AI-55) into equation (AI-21), the following expression is obtained.

$$\frac{1}{B_t} = \frac{2}{h_{BB}{}^2(t_p)} \int_{-\infty}^{\infty} h^2_{BB}\left(t + \frac{c_1}{2\pi}\right) \cos^2 2\pi f_{IF}\left(t + \frac{c'_0}{2\pi f_{IF}} + \frac{c_1}{2\pi}\right) dt \tag{AI-56}$$

or

$$\frac{1}{B_t} = \frac{2}{h_{BB}{}^2(t_p)} \int_{-\infty}^{\infty} h^2_{BB}\left(t + \frac{c_1}{2\pi}\right) \left[\tfrac{1}{2} + \tfrac{1}{2}\cos 4\pi f_{IF}\left(t + \frac{c'_0}{2\pi f_{IF}} + \frac{c_1}{2\pi}\right)\right] dt \tag{AI-57}$$

As a result of condition (g) of Appendix I-B-2, the second term in the integration is insignificant. Also, since the limits of integration are infinite, the expression simplifies (with a change of variable) to

$$\frac{1}{B_t} = \frac{1}{h_{BB}{}^2(t_p)} \int_{-\infty}^{\infty} h^2_{BB}(t)\, dt \tag{AI-58}$$

Thus, under the assumptions leading to equation (AI-39), the temporal bandwidth (B_t) is a system constant depending on $h^2_{BB}(t)$, the envelope of the square of the IF amplifier unit impulse response.

Using Parseval's formula given in equation (AI-9) to express the integral of equation (AI-58),

$$\frac{1}{B_t} = \frac{1}{h_{BB}{}^2(t_p)} \int_{-\infty}^{\infty} |H_{BB}(f)|^2\, df. \tag{AI-59}$$

The integral in this relation can be expressed in terms of H(f) as follows. First, integrating the magnitude squared of H(f) from equation (AI-36),

$$\int_{-\infty}^{\infty} |H(f)|^2 \, df = \int_{-\infty}^{\infty} \left| \frac{H_{BB}(f-f_{IF})}{2} + \frac{H_{BB}(f+f_{IF})}{2} \right|^2 df \qquad \text{(AI-60)}$$

As a result of condition (g) of Appendix I-B-2, this expression simplifies to

$$\int_{-\infty}^{\infty} |H(f)|^2 \, df = \int_{f_{IF}-(B'/2)}^{f_{IF}+(B'/2)} \left| \frac{H_{BB}(f-f_{IF})}{2} \right|^2 df + \int_{-f_{IF}-(B'/2)}^{-f_{IF}+(B'/2)} \left| \frac{H_{BB}(f+f_{IF})}{2} \right|^2 df \qquad \text{(AI-61)}$$

where B′ is the bandwidth of the frequency region of significant IF power gain defined in that condition. Changing variables yields

$$\int_{-\infty}^{\infty} |H(f)|^2 \, df = 2\int_{-(B/2)}^{(B/2)} \left| \frac{H_{BB}(f)}{2} \right|^2 df = \tfrac{1}{2}\int_{-\infty}^{\infty} |H_{BB}(f)|^2 \, df \qquad \text{(AI-62)}$$

Substituting equation (AI-62) into equation (AI-59) and changing the limits of integration to include only positive frequencies,

$$\frac{1}{B_t} = \frac{4}{h_{BB}^{\,2}(t_p)} \int_{0}^{\infty} |H(f)|^2 \, df \qquad \text{(AI-63)}$$

B_tB_{eff}, the product of the IF amplifier temporal bandwidth (B_t) and the IF amplifier effective random noise bandwidth (B_{eff}), can now be determined. From equation (AI-19) and the definition of G(f),

$$B_{eff} = \frac{1}{G(f_{IF})} \int_{0}^{\infty} G(f)\,df = \frac{1}{\dfrac{R_{in}}{R_{out}} |H(f_{IF})|^2} \int_{0}^{\infty} \frac{R_{in}}{R_{out}} |H(f)|^2 \, df \qquad \text{(AI-64)}$$

where R_{in} is the IF amplifier input resistance and R_{out} is the IF amplifier output resistance. Therefore, under the assumptions leading to equation (AI-39), the product B_tB_{eff} is given by

$$B_tB_{eff} = \frac{h_{BB}^{\,2}(t_p)}{4|H(f_{IF})|^2} \qquad \text{(AI-65)}$$

where $h_{BB}^2(t)$ is the envelope of the square of the IF amplifier unit impulse response, $h_{BB}^2(t_p)$ is this envelope evaluated at $t = t_p$ where it reaches its maximum, and $|H(f_{IF})|^2$ is the magnitude squared of the IF amplifier frequency characteristic, evaluated at the IF center frequency (f_{IF}).

Since the assumption leading to equation (AI-38) is equivalent to the assumption leading to equation (AI-17), then equation (AI-20) can be used to express the energy spectrum, $E_{IF}(f_{IF})$, at the spectrum analyzer IF amplifier input, evaluated at the IF center frequency (f_{IF}). This equation is

$$E_{IF}(f_{IF}) = \frac{p_{pk}}{2B_tB_{eff}G(f_{IF})} \tag{AI-20}$$

where p_{pk} is the maximum value of the IF output power during the transient, B_tB_{eff} is the product of the IF amplifier temporal bandwidth (B_t) and the IF amplifier random noise bandwidth (B_{eff}), and $G(f_{IF})$ is the IF amplifier power gain at the IF center frequency (f_{IF}). The B_tB_{eff} product can be calculated from equation (AI-65). Since, under the assumptions leading to equation (AI-39), this product is a system constant, it need be calculated only once for any particular spectrum analyzer.

(b) Relationship Between the IF Amplifier Output Transient and the IF Amplifier Input Energy Spectrum (Under Two Assumptions About the IF Amplifier Input Energy Spectrum and Assuming a Gaussian IF Amplifier Frequency Characteristic)

According to condition (g) of B.2 above, the spectrum analyzer IF amplifier unit impulse response, $h(t)$, can be approximated by

$$h(t) = h_{BB}(t)\cos(2\pi f_{IF}t) \tag{AI-66}$$

where $h_{BB}(t)$ contains only significant frequency components that are enough lower in frequency than the IF center frequency (f_{IF}) that the IF power gain (as a function of frequency) drops to a negligible level outside the frequency region defined by $f_{IF}-(B'/2) < f < f_{IF}+B'$, where B', the bandwidth of the frequency region of significant IF power gain, is related to the IF center frequency

(f_{IF}) by the inequality $B' \ll f_{IF}$. Thus, as a result of condition (g) of B.2,

$$H(f) = \frac{H_{BB}(f-f_{IF})}{2} + \frac{H_{BB}(f+f_{IF})}{2} \tag{AI-67}$$

where $H_{BB}(f)$ is the Fourier transform of $h_{BB}(t)$ and has significant magnitude only within the frequency region defined by $-(B'/2) \leqslant f \leqslant (B'/2)$ where B' is as defined above. Since $h_{BB}(t)$ must be real so that $h(t)$ will be real, then

$$|H_{BB}(f)| = |H^*_{BB}(-f)| \tag{AI-68}$$

It is seen that, if the IF amplifier frequency characteristic H(f), can be approximated by the Gaussian

$$H(f) = H_0 e^{-\left[\frac{\pi(f-f_{IF})}{k}\right]^2} + H_0 e^{-\left[\frac{\pi(f+f_{IF})}{k}\right]^2} \tag{AI-69}$$

where H_0 and k are real constants, then condition (g) of Appendix I-B-2 is satisfied provided k is large enough in magnitude. If such is the case, then from equations (AI-67) and (AI-69),

$$H_{BB}(f) = 2H_0 e^{-(\pi f/k)^2} \tag{AI-70}$$

Since $H_{BB}(f)$ is an even function of f, then

$$h_{BB}(t) = \int_{-\infty}^{\infty} H_{BB}(f)\, e^{j2\pi ft} df = 2\int_{0}^{\infty} H_{BB}(f) \cos 2\pi ftdt \tag{AI-71}$$

or

$$h_{BB}(t) = 4H_0 \int_0^{\infty} e^{-(\pi f/k)^2} \cos 2\pi ftdt \tag{AI-72}$$

With the aid of a table of integrals this equation can be reduced to

$$h_{BB}(t) = \frac{2k}{\sqrt{\pi}} H_0 e^{-(kt)^2} \tag{AI-73}$$

Squaring,

$$h_{BB}^2(t) = \frac{4k^2}{\pi} H_0^2 e^{-2(kt)^2} \tag{AI-74}$$

Thus, from equations (AI-65), (AI-69), and (AI-74),

$$B_t B_{eff} = \frac{\frac{4k^2}{\pi} H_0^2}{4H_0^2} = \frac{k^2}{\pi} \tag{AI-75}$$

This result can be expressed in terms of the 3 dB bandwidth as follows. From equation (AI-70),

$$H_{BB}(B/2) = 2H_0 e^{-(\pi B/2k)^2} \tag{AI-76}$$

and

$$H_{BB}(O) = 2H_0 \tag{AI-77}$$

By definition of the 3 dB bandwidth (B)

$$H_{BB}\left(\frac{B}{2}\right) = \frac{1}{\sqrt{2}} H_{BB}(O) \tag{AI-78}$$

Combining equations (AI-76), (AI-77), and (AI-78)

$$\frac{1}{\sqrt{2}} = e^{-(\pi B/2K)^2} \tag{AI-79}$$

$$-\ell n \sqrt{2} = - \frac{\pi B}{2k}^{\,2} \tag{AI-80}$$

$$k^2 = \frac{\pi^2 B^2}{2^2 \ell n \sqrt{2}} = \frac{\pi^2 B^2}{2\ell n 2} \tag{AI-81}$$

Substituting equation (AI-81) into equation (AI-75)

$$B_t B_{eff} = \frac{\pi}{2\ell n 2} B^2 \approx \left(\frac{3}{2}\right)^2 B^2 \tag{AI-82}$$

where B_t is the spectrum analyzer IF amplifier temporal bandwidth, B_{eff} is the spectrum analyzer IF amplifier random noise bandwidth, B is the spectrum analyzer IF amplifier 3 dB bandwidth, and the approximation $\pi/2\ell n2 \approx (3/2)^2$ is exact within 0.4%.

Thus, for such a frequency characteristic, under the two assumptions leading to equation (AI-39), one obtains from equation (AI-20),

$$E_{IF}(f_{IF}) = \frac{p_{pk}}{2\left(\frac{\pi}{2\ell n2}\right) B^2 G(f_{IF})} \approx \frac{p_{pk}}{2\left(\frac{3}{2}\right)^2 B^2 G(f_{IF})} \qquad \text{(AI-83)}$$

where $E_{IF}(f_{IF})$ is the energy spectrum at the spectrum analyzer IF amplifier input (evaluated at the IF center frequency, f_{IF}), p_{pk} is the maximum value of the IF output power during the transient, B is the spectrum analyzer IF amplifier 3 dB bandwidth, and $G(f_{IF})$ is the spectrum analyzer IF amplifier power gain at the IF center frequency (f_{IF}).

(c) Relationship Between the Energy Density Spectrum at the Test Site Antenna and the Energy Spectrum at the IF Amplifier Input

Since the relationship between the energy density spectrum at the test site antenna and the energy spectrum at the IF amplifier input does not depend on any radar emission assumptions, it is immediately obvious that the relation derived in C.2(b) can be used here. This relation is

$$E_{sD}(f_{RF}) = \frac{E_{IF}(f_{IF})}{AG_cG_1G_{RF}} \qquad \text{(AI-24)}$$

where $E_{sD}(f_{RF})$ is the energy density spectrum at the test site antenna input (evaluated at the frequency, f_{RF}, where the IF amplifier bandpass is effectively tuned during the pulse), $E_{IF}(f_{IF})$ is the energy spectrum at the IF amplifier input (evaluated at the IF center frequency, f_{IF}), A is the effective area of the test site antenna, G_c is the power gain of the cable (or waveguide) between the test site antenna and the spectrum analyzer input, G_1 is the attenuator gain between the test site antenna and the spectrum analyzer input, and G_{RF} is the power gain between the spectrum analyzer input and the IF amplifier input as a direct consequence of the real multiplicative constants mentioned in conditions (e) and (f) of Appendix I-B-2 and the impedance levels involved.

(d) Radar Emission Assumptions

It is obvious that the assumption made in obtaining equation (AI-38) is equivalent to the assumption made in obtaining equation (AI-17). Thus it can be concluded that the assumption made in obtaining equation (AI-38) is equivalent to radar emission assumption (1), repeated here from Appendix I-C-2-c for convenience.

Radar Emission Assumption (1)

$E_{sD}(f) \approx E_{sD}(f_{RF})$ for all frequencies (f) in the region defined by $f_{RF} - (B'/2) \leqslant f \leqslant f_{RF} + (B'/2)$, where $E_{sD}(f)$ is the energy density spectrum at the test site antenna input due to a single radar pulse, $E_{sD}(f_{RF})$ is $E_{sD}(f)$ evaluated at the RF frequency (f_{RF}), f_{RF} is the RF frequency where the IF bandpass is effectively tuned during the pulse, and B′ is the bandwidth of the frequency region of significant IF power gain.

From conditions (d), (e), and (f) of B.2, it is apparent that the assumption made in obtaining equation (AI-39) is equivalent to the following radar emission assumption.

Radar Emission Assumption (2):

$\phi(f) \approx K_0 + K_1 \times (f - f_{RF})$ for all frequencies (f) in the same frequency region as defined in radar emission assumption (1), where $\phi(f)$ is the phase angle of the Fourier transform of the electric field intensity at the test site antenna input due to a single radar pulse, K_0 and K_1 are coefficients of the first two terms of the Taylor series expansion of $\phi(f)$ about the point $f = f_{RF}$.

(e) Conclusion

Combining the results of C.3(a), (c), and (d), it is apparent that, under radar emission assumptions (1) and (2), the same conclusions are reached as were reached in C.2(d) under radar emission assumption (1), except that, under radar emission assumptions (1) and

(2), the B_tB_{eff} product is a system constant which can be calculated from equation (AI-65), as follows:

$$B_tB_{eff} = \frac{h_{BB}^2(t_p)}{4|H(f_{IF})|^2} \qquad \text{(AI-65)}$$

where $h_{BB}^2(t)$ is the envelope of the square of the spectrum analyzer IF amplifier unit impulse response, $h_{BB}^2(t_p)$ is this envelope evaluated at $t = t_p$ where it reaches its maximum, and $|H(f_{IF})|^2$ is the magnitude squared of the spectrum analyzer IF amplifier frequency characteristic, evaluated at the IF center frequency (f_{IF}). The result of this calculation, for a Gaussian IF frequency characteristic was shown in C.3 above to be

$$B_tB_{eff} = \frac{\pi}{2\ell n 2} B^2 \approx \left(\frac{3}{2}\right)^2 B^2 \qquad \text{(AI-82)}$$

where B is the spectrum analyzer IF amplifier 3 dB bandwidth.

Thus, under radar emission assumptions (1) and (2), the following conclusions are reached.

General Case:

$$P_{sD}(f_{RF}) = E_{sD}(f_{RF}) \times PRF = \frac{P_D}{B_tB_{eff}} \times PRF \qquad \text{(AI-27)}$$

Gaussian IF Frequency Characteristic:

$$P_{sD}(f_{RF}) = E_{sD}(f_{RF}) \times PRF = \frac{P_D}{\left(\frac{3}{2}\right)^2 B^2} \times PRF \qquad \text{(AI-84)}$$

In the above two equations, $P_{sD}(f_{RF})$ is the power density spectrum (power per unit area per unit frequency) at the test site antenna input (evaluated at f_{RF}), PRF is the pulse repetition frequency, $E_{sD}(f_{RF})$ is the energy density spectrum (energy per unit area per unit frequency) at the test site antenna input (evaluated at f_{RF}) f_{RF} is the RF frequency where the IF amplifier bandpass was effectively tuned during the radar pulse, B is the spectrum analyzer IF amplifier 3 dB bandwidth, P_D is the average CW power density (power per unit area) of frequency (f_{RF}) which, if applied to the test site antenna input, would produce the same vertical deflection on the spectrum analyzer (when the IF bandpass is effectively tuned to f_{RF}) as that produced by the radar pulse (when the IF bandpass was effectively tuned to f_{RF}), and $B_t B_{eff}$ is the product of the spectrum analyzer temporal and effective random noise bandwidths given by equation (AI-65) above.

Since, under radar emission assumptions (1) and (2), $B_t B_{eff}$ does not depend on $P_{sD}(f_{RF})$, then as a result of conditions (h) and (i) of B.2, only one point need be calibrated on the spectrum signature. Thus, the CW substitution data for a single graticule, obtained in step (6) of the test procedure presented in B.1, can be used to calculate P_D for that graticule and for the display center frequency (f_0), by using the relation

$$P_D = \frac{P_{sg}}{AG_cG_1} \qquad \text{(AI-28)}$$

where P_{sg} is the average CW power of frequency (f_0) substituted at the spectrum analyzer input, A is the effective area of the test site antenna, G_c is the power gain of the cable (or waveguide) between the test site antenna and the spectrum analyzer input, and G_1 is the attenuator power gain between the test site antenna and the spectrum analyzer input. By using all the CW substitution data obtained in step (6) of the test procedure described in B.1, one can experimentally calibrate each graticule line at the center of the screen and the center horizontal graticule at both ends of the screen. Thus a check is provided on how well the spectrum analyzer meets conditions (h) and (i) of B.2. If it is found that these conditions are poorly met, then the various calibrated points can be used to interpolate to some extent intermediate points.

4. Determination of which Radars Meet Radar Emission Assumptions (1) and (2)

(a) Pulse Radars

Rectangular Waveshape:

For a rectangular pulse shape, the theoretical electric field intensity, y(t), at the test site antenna is given by

$$\begin{aligned} y(t) &= A\cos 2\pi f_0 t \qquad & |t| \leqslant \frac{\tau}{2} \\ &= 0 & |t| > \frac{\tau}{2} \end{aligned} \tag{AI-85}$$

where A is the amplitude of y(t), f_0 is the RF carrier frequency, and τ is the pulse duration (time between half power points). Assuming $(1/\tau) \ll f_0$, then y(t) can be factored into a baseband factor, $y_{BB}(t)$, and a carrier factor:

$$y(t) = y_{BB}(t)\cos 2\pi f_0 t \tag{AI-86}$$

where $y_{BB}(t)$ contains only significant frequency components which are enough lower than f_0 that the Fourier transform of $y_{BB}(t)$ has significant magnitude only in the frequency region defined by $-(B''/2) \leqslant f \leqslant (B''/2)$, where B″ equals the width of this frequency region and is related to the carrier frequency (f_0) by the inequality, $B'' \ll f_0$. From equations (AI-85) and (AI-86), we get

$$\begin{aligned} y_{BB}(t) &= A \qquad & |t| \leqslant \frac{\tau}{2} \\ &= 0 & |t| > \frac{\tau}{2} \end{aligned} \tag{AI-87}$$

Thus,

$$Y_{BB}(f) = \int_{-(\tau/2)}^{(\tau/2)} Ae^{-j2\pi ft}\,dt = \frac{A}{-j2\pi f}\left[e^{-j2\pi ft}\right]_{-(\tau/2)}^{(\tau/2)} \tag{AI-88}$$

$$Y_{BB}(f) = \frac{A}{\pi f} \sin\pi f\tau = A\tau \frac{\sin\pi f\tau}{\pi f\tau} \tag{AI-89}$$

From equation (AI-86),

$$Y(f) = \int_{-\infty}^{\infty} y_{BB}(t) \left[\frac{e^{j2\pi f_0 t} - e^{-j2\pi f_0 t}}{2}\right] e^{-j2\pi ft}\, dt \tag{AI-90}$$

$$Y(f) = ½\int_{-\infty}^{\infty} y_{BB}(t) e^{-j2\pi(f-f_0)t} dt - ½\int_{-\infty}^{\infty} y_{BB}(t) e^{-2\pi(f+f_0)t} dt \tag{AI-91}$$

The integrals in this equation are easily recognized, and Y(f) can be written as

$$Y(f) = \frac{Y_{BB}(f-f_0)}{2} + \frac{Y_{BB}(f+f_0)}{2} \tag{AI-92}$$

The two-sided, 1-ohm energy density, $|Y(f)|^2$ is given by

$$|Y(f)|^2 = \left|\frac{Y_{BB}(f-f_0)}{2} + \frac{Y_{BB}(f+f_0)}{2}\right|^2 \tag{AI-93}$$

Because of the bandlimited nature of $Y_{BB}(f)$, mentioned above, this equation can be written

$$|Y(f)|^2 = \frac{|Y_{BB}(f-f_0)|^2}{4} + \frac{|Y_{BB}(f+f_0)|^2}{4} \tag{AI-94}$$

The energy density, $E_{sD}(f)$, at the test site antenna is twice the positive frequency portion of the two sided, 1-ohm energy density divided by the resistance of free space. Thus,

$$E_{sD}(f) = \frac{|Y_{BB}(f-f_0)|^2}{2R_s} \tag{AI-95}$$

where R_s is the resistance of free space (120π ohms). Substituting equation (AI-89) into equation (AI-95), the following result is obtained:

$$E_{sD}(f) = \frac{A^2\tau^2}{2R_s} \frac{\sin^2[\pi(f-f_0)\tau]}{[\pi(f-f_0)\tau]^2} . \tag{AI-96}$$

The instantaneous power density at the test site antenna is the square of the electric field intensity divided by the resistance of free space. Thus, from equation (AI-85), the greatest one-cycle average, $\hat{P}_D$, of the instantaneous power density is

$$\hat{P}_D = \frac{A^2}{2R} \tag{AI-97}$$

where $\hat{P}_D$ will be referred to as the peak power density (power per unit area). Thus, the theoretical energy density spectrum at the test site antenna is given by

$$E_{sD}(f) = \hat{P}_D\tau^2 \frac{\sin^2[\pi(f-f_0)\tau]}{[\pi(f-f_0)\tau]^2} \tag{AI-98}$$

From equations (AI-89) and (AI-92), the theoretical Fourier transform of the electric field intensity at the test site antenna is given by,

$$Y(f) = \frac{A\tau}{2} \frac{\sin[\pi(f-f_0)\tau]}{\pi(f-f_0)\tau} + \frac{A\tau}{2} \frac{\sin[\pi(f+f_0)\tau]}{\pi(f+f_0)\tau} \tag{AI-99}$$

It is apparent from equations (AI-98) and (AI-99) that the meeting of radar emission assumptions (1) and (2) is assured by the requirement in step (3) of the test procedure given in Appendix I-B-1 which states that

$$B < \frac{1}{10\tau} \tag{AI-2}$$

for fixed frequency pulses, where τ is the pulse duration (time between half power points).

In a similar fashion, the theoretical energy density spectrum, $E_{sD}(f)$, at the test site antenna and the Fourier transform, $Y(f)$, of the theoretical electric field intensity at the test site antenna have been derived for cosine, triangular, cosine squared, and Gaussian waveshapes. The results are summarized below. It is apparent from these results that the meeting of radar emission assumptions (1) and (2) is similarly assured by the IF bandwidth requirement in step (3) of the test procedure given in Appendix I-B-1.

Cosine Waveshape:

$$y(t) = A\cos\left(\frac{\pi t}{b}\right)\cos(2\pi f_0 t) \qquad |t| \leqslant \frac{b}{2}$$

$$= 0 \qquad |t| > \frac{b}{2} \tag{AI-100}$$

$$b = 2\tau \tag{AI-101}$$

$$E_{sD}(f) = \frac{4\hat{P}_D b^2}{\pi^2}\left|\frac{\cos[\pi(f-f_0)b]}{1-4(f-f_0)^2b^2}\right|^2 \tag{AI-102}$$

$$Y(f) = \frac{Ab}{4}\,\frac{\sin\left(\pi(f-f_0)b-\frac{\pi}{2}\right)}{\pi(f-f_0)b-\frac{\pi}{2}} + \frac{Ab}{4}\,\frac{\sin\left(\pi(f-f_0)b+\frac{\pi}{2}\right)}{\pi(f-f_0)b+\frac{\pi}{2}}$$

$$+\frac{Ab}{4}\,\frac{\left(\sin\ \pi(f+f_0)b-\frac{\pi}{2}\right)}{\pi(f+f_0)b-\frac{\pi}{2}} + \frac{Ab}{4}\,\frac{\sin\left(\pi(f+f_0)b+\frac{\pi}{2}\right)}{\pi(f+f_0)b+\frac{\pi}{2}} \tag{AI-103}$$

Triangular Waveshape:

$$y(t) = A\left(1+\frac{2t}{b}\right)\cos(2\pi f_0 t) \qquad -\frac{b}{2} \leqslant t \leqslant 0$$

$$= A\left(1-\frac{2t}{b}\right)\cos(2\pi f_0 t) \qquad 0 \leqslant t \leqslant \frac{b}{2} \tag{AI-104}$$

$$= 0 \qquad |t| > \frac{b}{2}$$

$$b = \frac{2\tau}{(2 - \sqrt{2}} \tag{AI-105}$$

$$E_{sD}(f) = \frac{\hat{P}_D b^2}{4} \frac{\sin^4\left[\pi(f-f_0)\frac{b}{2}\right]}{\left[\pi(f-f_0)\frac{b}{2}\right]^4} \tag{AI-106}$$

$$Y(f) = \frac{Ab}{4} \frac{\sin^2\left[\pi(f-f_0)\frac{b}{2}\right]}{\left[\pi(f-f_0)\frac{b}{2}\right]^2} + \frac{Ab}{4} \frac{\sin^2\left[\pi(f+f_0)\frac{b}{2}\right]}{\left[\pi(f+f_0)\frac{b}{2}\right]^2} \tag{AI-107}$$

Cosine Squared Waveshape:

$$y(t) = A\cos^2\left(\frac{\pi t}{b}\right)\cos 2\pi f_0 t \qquad |t| \leqslant \frac{b}{2}$$

$$= 0 \qquad |t| > \frac{b}{2} \tag{AI-108}$$

$$b = \frac{\pi\tau}{\cos^{-1}(\sqrt{2} - 1)} \tag{AI-109}$$

$$E_{sD}(f) = \frac{\hat{P}_D b^2}{4} \left|\frac{\sin[\pi(f-f_0)b]}{\pi(f-f_0)b[1-(f-f_0)^2b^2]}\right|^2 \tag{AI-110}$$

$$Y(f) = \frac{Ab}{4} \frac{\sin[\pi(f-f_0)b]}{\pi(f-f_0)b} + \frac{Ab}{8} \frac{\sin[\pi(f-f_0)b-\pi]}{\pi(f-f_0)b-\pi}$$

$$+ \frac{Ab}{8} \frac{\sin[\pi(f-f_0)b+\pi]}{\pi(f-f_0)b+\pi} + \frac{Ab}{4} \frac{\sin[\pi(f+f_0)b]}{\pi(f+f_0)b}$$

$$+\frac{Ab}{8}\,\frac{\sin[\pi(f+f_0)-\pi]}{\pi(f+f_0)-\pi}+\frac{Ab}{8}\,\frac{\sin[\pi(f+f_0)b+\pi]}{\pi(f+f_0)b+\pi} \tag{AI-111}$$

Gaussian Waveshape (Non-realizable):

$$y(t) = Ae^{-k^2t^2} \tag{AI-112}$$

$$k^2 = \frac{2\ell n2}{\tau^2} \tag{AI-113}$$

$$E_{sD}(f) = \frac{\pi\hat{P}_D}{k^2}\, e^{-\frac{2\pi^2}{k^2}(f-f_0)^2} \tag{AI-114}$$

$$Y(f) = \frac{A\sqrt{\pi}}{2k}\, e^{-\frac{\pi^2}{k^2}(f-f_0)^2} + \frac{A\sqrt{\pi}}{2K}\, e^{-\frac{\pi^2}{k^2}(f+f_0)^2} \tag{AI-115}$$

(b) Pulse Doppler Radars

As a result of the requirement in step (3) of the test procedure given in B.1, the meeting of radar emission assumptions (1) and (2) is similarly assured for pulse doppler radars with rectangular, cosine, triangular, cosine squared, or Gaussian waveshapes. Pulse doppler radars, though, are less likely to meet condition (b) of B.2, because of their high pulse repetition frequencies.

5. Conclusion – Recommended Calibration Technique

Every type of radar investigated in C.5 which met radar emission assumption (1) also met radar emission assumption (2). Thus, the calibration techniques developed in C.3, under radar emission assumptions (1) and (2), are recommended. These techniques (one for a general spectrum analyzer IF amplifier frequency characteristic and one for a Gaussian spectrum analyzer IF amplifier frequency characteristic) are much easier to use than the technique developed in C.2, under radar emission assumption (1) alone.

Throughout C, a compatible system of units has been assumed. This fact is indicated by the use of unity proportionality constants in all equations. The power density spectrum levels specified in MIL-STD-469 in the definition of radar emission bandwidth are incompatible, though. In order to obtain the power density spectrum, $P_{sD}(f_{RF})$, at the test site antenna in mw/kHz/m^2 it is necessary to multiply the results from equation (AI-27) or (AI-84) of C.3 by 10^3. Thus in order to obtain the power density spectrum, $P_{sD}(f_{RF})$, at the test site antenna in dB with respect to 1 mw/kHz/m^2, under radar emission assumptions (1) and (2), one of the following equations must be used.

General Case:

$$P_{sD}(\text{dBm/kHz/m}^2) = P_D(\text{dBm/m}^2) - 10\log\left[B_t(\text{Hz})B_{eff}(\text{Hz})\right] + 10\log\frac{\text{PRF(Hz)}}{1\text{ Hz}} + 30\text{ dB} \quad \text{(AI-116)}$$

Gaussian IF Frequency Characteristic:

$$P_{sD}(\text{dBm/kHz/m}^2) = P_D(\text{dBm/m}^2) - 20\log\left(\frac{3}{2}B\right) + 10\log\frac{\text{PRF(Hz)}}{1\text{ Hz}} + 30\text{ dB} \quad \text{(AI-117)}$$

In the above equations, P_{sD}(dBm/kHz/m^2) and P_D(dBm/m^2) are measured at the RF frequency, f_{RF}, where the power density spectrum is to be calibrated. P_D(dBm/m^2) is the average CW power density (measured in dB with respect to 1 mw/m^2) which, if applied to the test site antenna input, would produce the same vertical deflection on the spectrum analyzer (when the IF bandpass is effectively tuned to f_{RF}) as that produced by the radar pulse (when the IF bandpass was effectively tuned to f_{RF}). PRF(Hz) is the pulse repetition frequency (measured in Hz), B is the spectrum analyzer IF amplifier 3 dB bandwidth (measured in Hz), and B_t(Hz)B_{eff}(Hz) is the product (measured in Hz^2) of the spectrum analyzer IF amplifier temporal bandwidth (B_t) and the spectrum analyzer IF amplifier effective random noise bandwidth (B_{eff}).

The $B_t B_{eff}$ product can be calculated in the appropriate units from equation (AI-65) as follows:

$$B_t(\text{Hz})\, B_{eff}(\text{Hz}) = \frac{h_{BB}^2(t_p)}{4|H(f_{IF})|^2} \qquad \text{(AI-65)}$$

where $h_{BB}^2(t)$ is the envelope of the square of the spectrum analyzer IF amplifier unit impulse response, $h_{BB}^2(t_p)$ is this envelope evaluated at $t = t_p$ where it reaches its maximum, and $|H(f_{IF})|^2$ is the magnitude squared of the spectrum analyzer IF amplifier frequency characteristic, evaluated at the IF center frequency.

The power density, $P_D(\text{dBm/m}^2)$, can be calculated in the appropriate units from the equation

$$P_D(\text{dBm/m}^2) = P_{sg}(\text{dBm}) - 10 \log \frac{A(\text{m}^2)}{1\text{m}^2} + A_c(\text{dB}) + A_1(\text{dB}) \qquad \text{(AI-118)}$$

where $P_D(\text{dBm/m}^2) = P_{sg}(\text{dBm})$ are measured at the RF frequency (f_{RF}) where the power density spectrum is to be calibrated. P_{sg} (dBm) is the average CW power (measured in dBm) substituted at the spectrum analyzer input; A_c(dB) is the power loss (measured in dB) of the cable (or waveguide) between the test site antenna and the spectrum analyzer input; A_1(dB) is the attenuator power loss (measured in dB) between the test site antenna and the spectrum analyzer input; and $A(\text{m}^2)$ is the effective area (measured in square meters) of the test site antenna.

The effective area, $A(\text{m}^2)$, of the test site antenna can be calculated in the appropriate units from the equation

$$A(\text{m}^2) = \frac{[\lambda(\text{m})]^2 G_T}{4\pi} \qquad \text{(AI-119)}$$

where $\lambda(\text{m})$ is the wavelength (measured in meters) of the radar emission and G_T is the power gain over isotropic (measured as a unitless ratio) of the test site antenna. Usually it is more convenient to calculate 10 log $(A(m^2)/1m^2)$ directly from the equation

$$10 \log \frac{A(m^2)}{1m^2} = G_T(dB) - 20 \log \frac{f(MHz)}{1\ MHz} + 38.55\ dB \qquad \text{(AI-120)}$$

where f(MHz) is the frequency (measured in MHz) of the radar emission and $G_T(dB)$ is the power gain over isotropic (measured in dB) of the test site antenna.

Finally, since the power density spectrum levels specified in MIL-STD-469 are corrected to a distance of 1850 meters, the result obtained from equation (AI-116) or (AI-117) must be corrected to that distance by the following equation:

$$P'_{sD}(dBm/kHz/m^2) = P_{sD}(dBm/kHz/m^2) + 20 \log \frac{d_0(m)}{1850\ m} \qquad \text{(AI-121)}$$

where $P'_{sD}(dBm/kHz/m^2)$ is the power density spectrum level corrected to a distance of 1850 meters, $P_{sD}(dBm/kHz/m^2)$ is the power density spectrum level at the test site antenna, and $d_0(m)$ is the distance between the transmitting antenna and the test site antenna.

*In the case of the test site antenna feeding a waveguide, the output signal from the antenna is a field intensity.

**By zero level linear phase shift is meant a linear phase shift which has zero value if extrapolated to zero frequency. In any case, the total phase shift (constant plus zero level linear) at any positive frequency (f) will be the negative of the total phase shift at the corresponding negative frequency (-f).

†In the case of a waveguide feeding the spectrum analyzer, the incoming signal to the first mixer is a field intensity and the resulting signal is a voltage. Thus the mixer acts as a transducer as well as a frequency shifter.

*As a result of conditions (h) and (i) of Appendix I-B-2, these maximum voltages are recorded in terms of the associated maximum powers, the units being dB with respect to some arbitrary reference.

*B.O. Pierce, *A Short Table of Integrals*, 4th ed., c. 1956, Ginn and Company, New York, N.Y., page 69.

